CW01184202

EVE OF DESTRUCTION

COLONEL JOHN HUGHES-WILSON

EVE OF DESTRUCTION

THE INSIDE STORY OF OUR DANGEROUS NUCLEAR WORLD

First published in the UK by John Blake Publishing
an imprint of Bonnier Books UK
The Plaza,
535 Kings Road,
London SW10 0SZ
Owned by Bonnier Books
Sveavägen 56, Stockholm, Sweden

www.facebook.com/johnblakebooks
twitter.com/jblakebooks

First published in hardback in 2021

Hardback ISBN: 978-1-78946-337-8
Ebook ISBN: 978-1-78946-338-5
Audio book ISBN: 978-1-78946-347-7

All rights reserved. No part of this publication may be reproduced, stored in a retrieval system, or transmitted in any form or by any means, without the prior permission in writing of the publisher, nor be otherwise circulated in any form of binding or cover other than that in which it is published and without a similar condition including this condition being imposed on the subsequent purchaser.

British Library Cataloguing-in-Publication Data:

A catalogue record for this book is available from the British Library.

Design by www.envydesign.co.uk

Printed and bound in Great Britain by Clays Ltd, Elcograf S.p.A.

1 3 5 7 9 10 8 6 4 2

Text copyright © John Hughes-Wilson 2021

The right of John Hughes-Wilson to be identified as the author of this work has been asserted by him in accordance with the Copyright, Designs and Patents Act 1988.

Extracts from the author's earlier work, *A Brief History of the Cold War* (London, Robinson, 2006), are reproduced by permission of Little, Brown Book Group.

MIX
Paper from responsible sources
FSC® C018072

Every reasonable effort has been made to trace copyright-holders of material reproduced in this book, but if any have been inadvertently overlooked the publishers would be glad to hear from them.

John Blake Publishing is an imprint of Bonnier Books UK
www.bonnierbooks.co.uk

This book is dedicated to Andrew Hayward, whose idea it was. One of the silverbacks of the London publishing scene and a literary agent par excellence.

CONTENTS

Foreword		ix
Introduction		xi
1	Atoms	1
2	Ruthless Rays: Radiation – Silent but Deadly	17
3	Oops! Some Early Accidents	29
4	Some Early Military Near-misses	43
5	Balls-up at Bikini	59
6	The French Foul-up	71
7	Stop Digging! Civilian Engineering	79
8	The Nuclear Navy	85
9	All at Sea: Russia's Terrible Record	95
10	The *Kursk* Catastrophe	113
11	Manna from Heaven? Lost Bombs	123
12	Fireworks Can Be Dangerous	139
13	Radiation Is Really Bad for Your Health	153
14	You Can't Brush Nuclear Under the Carpet	169

15	Britain's Dirty Secret: The Windscale Fire	187
16	The World Held Its Breath: Cuba 1962	201
17	The Real Cuban Missile Crisis – At Sea	215
18	Exercise Able Archer 83	223
19	Some Civilian Nuclear Disasters	235
20	Chernobyl – The Worst so Far	253
21	Eastern Enigmas: China, India, Pakistan – And Israel	267
22	Japan in the Dock	287
23	Japan's Nuclear Nightmare: Fukushima	295
24	False Alarms	309
25	The Sum of All Fears: Terrorists with a Nuke	323
26	Future Uncertain? Summary and Conclusions	341

Appendix: Nuclear-armed Countries with Weapons-grade Highly Enriched Uranium, and Countries with Nuclear Power	349
Notes and Sources	351
Index	373

FOREWORD

This book came as a result of a suggestion by my agent. At first I was none too keen; there seemed to be an awful lot of books on the market covering nuclear accidents and disasters already. However, after forty years in the business, Andy Hayward knows the publishing world like few others, and as I researched around I began to see his point. Further encouragement came from Toby Buchan of John Blake Publishing, who had the foresight to see the commercial potential in my book about the assassination of President John F Kennedy, published back in 2013.

It is true that there are a lot of 'nuclear' books out there, but most of them tend to concentrate on a single issue. There are literally dozens of books about the accidents at Chernobyl, Three Mile Island and Fukushima, but very few covering the wider subject of nuclear cock-ups over the years; and even a cursory trawl showed that there have been a lot more accidents than we realise.

The more I delved, the more I found; and the more I found, the more I began to realise how neglected the saga of nuclear mishaps, blunders and sometimes rank incompetence has been over the years. Some of the things I discovered were genuinely shocking – and on some occasions frightening. Never mind Kennedy and Cuba: we owe a massive debt to men like Commodore Arkhipov and Colonel Petrov. It can be argued that their common sense and quick thinking on the spot literally saved the world. I had no idea just how close we had really come to Armageddon.

My own interest is based on personal experience, having been an intelligence officer at Supreme Headquarters Allied Powers Europe (SHAPE) during exercise Able Archer in 1983. Some of the material used in the chapters on Cuba and Able Archer 83 is drawn from my close personal knowledge, suitably updated. I may have slept as an exhausted officer cadet through the Cuban Missile Crisis of 1962, but I was well aware in 1983 that something historic and momentous was going on in the SHAPE Operations Room around me. We all were.

From my close personal knowledge, I have used the material in the Cuba and Able Archer chapters of *Intelligence Blunders* and *On Intelligence*, suitably updated.

This book is therefore the distillation of a professional interest in the subject, allied with a deep digging into the stories of what really went on with our risky nuclear toys, be they weapons or peaceful power stations. All are potentially very, very dangerous; and we humans do have a worrying record of mistakes.

This book serves as an exposure and a warning.

John Hughes-Wilson
Cyprus, 2021

INTRODUCTION

'We have grasped the mystery of the atom and rejected the Sermon on the Mount.'
– GENERAL OMAR N BRADLEY

Nuclear weapons are terrifying. Opinions about their development and continued existence vary enormously, but they scare the hell out of me – a trained nuclear, biological and chemical warfare specialist.

No one who has not experienced the effects of chemical weapons on the human body, or witnessed a nuclear explosion, can have any real understanding of their deadly impact. Survivors' accounts from the two atom bombs dropped on Japan in 1945 are more than frightening. Since then, scientific research has uncovered new horrors and an even more devastating everyday nuclear reality, unknown or ignored by most people as just too unpleasant to think about.

But, however hard it is to contemplate, we should not pretend that a major nuclear event is never going to happen. Nuclear disasters are a genuine reality, and we need to understand their consequences if we are to have any appreciation of the true dangers of nuclear weapons, whether they are detonated by some fanatical terrorist, a rogue state or even by accident – of which there have been far too many. While the world has avoided a nuclear war, the number of mishaps, as chronicled in this book, is of grave concern. Nuclear accidents are two a penny.

We really do need to sit up and take notice of the dangers of a nuclear war – which is becoming increasingly more likely as the years go by – as well as the all-too-frequent civilian nuclear emergencies that have threatened us since the very first nuclear accident, long before the first atom bomb.

While the effects of conventional explosions are well known, even by civilians in this age of terrorist IEDs – improvised explosive devices – we need to understand that nuclear weapons are different. They are much more than just a very big bang, as would be generated by lots of TNT explosive. Nuclear explosions are emphatically *not* like big conventional explosions. They are much worse. Nuclear weapons don't just blow you to pieces. They can vaporise, kill, wound and maim in four separate ways – and we need to understand just how deadly these threats are.

The first thing that anyone unlucky enough to be in the vicinity of a nuclear detonation would notice is the intense double pulse of light, often described as 'brighter than a thousand suns', a phrase based on a line from the *Bhagavad Gita*, an ancient Hindu scripture. That flash will fry your eyeballs and even burn your shadow on the wall like a grim photographic negative or eerie memorial.

Then comes the intense heat, a pulse of tens of millions of degrees, racing out from ground zero's luminous fireball at almost the speed of light. The heat is so devastating that it vaporises wood and even bricks, let alone soft tissue, flesh and bone. Human beings combust to ash in a second. Even glass can melt and catch fire.

Third comes the all-powerful explosive blast wave, radiating out from ground zero at over 4,800 kilometres per hour (3,000mph), literally 'faster than a speeding bullet', destroying everything in its path. Buildings are ripped apart; cars, people and animals are blown away by a ferocious gale, a hundred times worse than the worst tornado. Everything standing is vulnerable to the blast. At Hiroshima, only the strongest concrete structures survived and then as twisted, broken shells. Today's buildings actually increase our nuclear vulnerability, thanks to the vanity of modern architects. In a nuclear attack, the name of London's Shard would prove all too appropriate as a million razor-sharp fragments of flying glass would scatter for miles around.

Last and most insidious of all, after the heat and the blast comes radiation – invisible, silent and deadly: the surreptitious killer. Radiation is not a phenomenon unique to nuclear explosions or accidents. It occurs naturally on our planet in a number of different forms, but the intensity and extent of the radiation created by a man-made nuclear event is on an altogether different scale. Worst of all, because you can't see, hear or feel radiation, it attacks you without you knowing. It permeates the ground and the buildings around you, it becomes absorbed deep in the soil and turns everything it touches into a nuclear hazard. It may not always glow in the dark, but lurking in every single thing, animate or inanimate, that has been exposed to a nuclear pulse, radiation waits, pulsing out its invisible rays to maim and kill, often long after the nuclear event.

Nuclear radiation doesn't go away. Some of its hazards can be short-lived, but most will remain a long-term problem as it lies dormant, contaminating the soil and area downwind of the explosion, often for thousands of kilometres. Years after the acute first effects of it have subsided, residual radiation will continue to produce a wide range of physical problems for living human beings, from leukaemia and other cancers to malformed babies. So, if the light, heat or blast don't get you, then the radiation will – or it will make your neighbourhood uninhabitable for a long time. Nuclear weapons are very different from conventional explosives, in almost every way.

US President Harry S Truman did not understand the full potential of nuclear weapons when he ordered the first atom bombs to be dropped on Japan back in 1945, in what proved to be the final year of World War II. As far as he was concerned, he was only authorising a new, monster explosion. But at that time, only a handful of scientists, such as J Robert Oppenheimer, really knew what they were unleashing on to an unsuspecting world. As Truman himself recorded in his diary for 25 July that year:

> We have discovered the most terrible bomb in the history of the world. It may be the fire destruction prophesied in the Euphrates Valley Era, after Noah and his fabulous Ark.
>
> Anyway we 'think' we have found the way to cause a disintegration of the atom. An experiment in the New Mexico desert was startling – to put it mildly . . . The explosion was visible for more than 200 miles [320 kilometres] and audible for 40 miles [65 kilometres] and more . . .
>
> It is certainly a good thing for the world that Hitler's crowd or Stalin's did not discover this atomic bomb. It seems to be the most terrible thing ever discovered, but it can be made the most useful.

Truman had no qualms about using the bomb to bring about a Japanese surrender and save the American lives that would have been sacrificed in the invasion of the Japanese home islands. Planners estimated that such an invasion, dubbed 'Operation Olympic', would result in around a million casualties. The US President believed he was only witnessing a step change in humanity's brutal arsenal of warfare. Just a bigger bomb? Now we know differently.

Since 1945 the problem has grown, steadily, stealthily and relentlessly. Following the end of World War II, there have been no more nuclear attacks. Although nuclear weapons arsenals have multiplied, so far they have lain like dragons sleeping in their caves.

The problem is that 'nuclear' as a potential threat has not gone away. There have been a disturbing number of nuclear accidents and close shaves, both military and civilian. More serious still is the all-too-real threat of nuclear weapons falling into the hands of Islamic fanatics or like-minded terrorist groups. Nuclear proliferation, in all its forms, continues to be a deadly serious and growing danger.

Should al-Qa'ida, ISIS or any other fanatical jihadi group ever get their hands on a Pakistani nuclear weapon, then they would have a weapon with a yield of up to 40 kilotons. One kiloton is the blast equivalent of 1,000 tonnes of TNT. If smuggled on board a ship into the Pool of London or the New York docks, a 40-kiloton device would obliterate an area of over fifty square kilometres – enough to make the 9/11 World Trade Center attacks look like a small-scale rehearsal and to render Manhattan or London uninhabitable for a generation.

Aside from nuclear war, this form of terrorism is perhaps the most potent and present danger that we face today. It has been the subject of many fictional books and Hollywood movies, but we cannot afford to dismiss it as fanciful, much as we might wish to. We already have plenty of evidence to show just how fragile our control is over nuclear weapons, their storage, facilities and security. Yet there are even worse nuclear-related dangers than a terrorist threat.

The risk of a devastating nuclear accident is far more likely. Human beings control these lethal devices, and human beings are very, very fallible. Nuclear accidents like the one that took place in Chernobyl in 1986 are more likely than a nuclear war or even an incident of nuclear terrorism.

At the heart of the problem is the fact that we cannot 'disinvent' nuclear weapons, however much we might like to, just as a worried papacy could not disinvent the medieval crossbow or in the early nineteenth century, the British government could not ban the new-fangled torpedo on the grounds that a weaker navy might use it to gain an advantage over a strong one, such as Britain's. Like a genie, a new weapon, once out of the bottle, cannot be coaxed back inside.

Since 1945, there have been hundreds of recorded nuclear accidents, civilian and military, many of which could have caused as much, if not more, devastation than any Hiroshima bomb. The sad truth is that the record on nuclear mishaps over the last seventy-five years does not bode well for our future. They stand as a grim warning that if we mean to survive, then we really must exercise tighter control over nuclear security.

The sobering conclusion is that one day, inevitably, a nuclear war, a terrorist maniac or a bungled accident will cause unimaginable devastation. Given humanity's track record, it can only be a matter of time.

We have had plenty of warning.

CHAPTER 1

ATOMS

'All things are made of atoms – little particles that move around in perpetual motion, attracting each other when they are a little distance apart, but repelling upon being squeezed into one another.'
– **RICHARD P FEYNMAN, PHYSICIST**

The atom is the basic unit of matter. It is the smallest thing that can have a distinct chemical property. There are many different types of atom, each with its own name, atomic mass and size, depending on its host material.

The Greek philosopher Democritus first coined the word 'atom' in around 400 BC; it comes from the Greek 'atomos' (ἀτόμος), or 'indivisible'. Democritus had a theory that everything was made up of small, invisible balls of matter. As no one could verify this at the time, let alone see an atom, the idea remained merely the subject of speculative philosophical debate until about 1650, when modern chemistry began to emerge. Truly, we owe much to the insatiable curiosity that characterised the Enlightenment.

By 1777, a French nobleman called Antoine Lavoisier, since dubbed the 'father of modern chemistry', had made a number of important discoveries, among them the existence of oxygen, and had realised that matter was composed of various different elements. Lavoisier proposed a definition of an element, and understood that matter consisted of indivisible particles – atoms. His reward, despite his discoveries and his record as a distinguished member of the French Academy of Sciences, was to be guillotined during the Revolution because he was an aristocrat and therefore an enemy of 'the people'.

The British picked up on Lavoisier's work. By 1803, a Quaker philosopher called John Dalton had begun working on a theory that the various elements were composed of small solid balls that could combine together to form compounds. Dalton's notebooks never made clear how he came to his conclusions, but they were far-reaching in their impact on chemistry. He confirmed that elements must be composed of extremely small particles, called atoms, and argued that:

1. Atoms of a given element are identical in size, mass and other physical properties.
2. Atoms of different elements differ in size, mass and other properties.
3. Atoms cannot be destroyed.
4. Atoms of different elements combine to form chemical compounds.
5. In chemical reactions, atoms are rearranged by combination or separation.

Dalton's work inevitably spawned questions in turn, notably: What was the real nature of matter? And could atoms be broken or split apart?

During the nineteenth century, various scientists began to study the composition of matter and to identify and codify the elements. By 1869, a Russian scientist named Dmitri Mendeleev (whose scruffy beard, wild eyes and long hair have prompted comparisons with Rasputin) had determined to find a better way of organising the study of chemistry. This led to his development of the periodic law – stating that the chemical and physical properties of elements recur in an ordered and predictable pattern when they are arranged in terms of increasing atomic number. He illustrated this by creating the periodic table, one of the most iconic and instantly recognisable symbols in science. (The German chemist Lothar Meyer independently devised his own version of the periodic law and table at around the same time.) More importantly, Mendeleev used his periodic table to predict the existence and properties of chemical elements then unknown to science, but which were subsequently discovered; his place in the history of science was assured. One of the elements that attracted particular interest was uranium.

Uranium's existence had been known for a long time. In 1789, the German chemist Martin Klaproth was studying the mineral pitchblende (now generally known as uraninite), a black, dense ore found in the Congo and Canada. (It is very slightly radioactive, owing to its natural traces of uranium and also contains a small amount of radium as a naturally radioactive by-product of uranium decay.)

By dissolving pitchblende in nitric acid, Klaproth produced a yellow substance that he concluded was a new element; he named it uranium, after the planet Uranus. By the 1890s, scientists had begun to understand that uranium was a very strange element, requiring deeper study. French physicist Henri Becquerel demonstrated that uranium and radium gave off mysterious rays, and it was while studying radiation from radium that Paul Villard discovered gamma rays in 1900. A few years earlier, in Germany, Wilhelm Röntgen had accidentally discovered X-rays after he noticed that a luminescent screen left near a discharge tube had been affected by a mysterious radiation source. Husband and wife Pierre and Marie Curie further developed these theories of radiation. By the end of the nineteenth century, the British Army had even used mobile X-ray units in a number of overseas campaigns, including the War of the Sudan and the Boer War, to locate bullets and shrapnel in wounded soldiers. The pace of scientific progress in matters nuclear was accelerating.

New Zealand-born physicist Ernest Rutherford made the study of atomic physics his Nobel Prize-winning life's work. During the 1890s, he studied at the Cavendish Laboratory at the University of Cambridge with a distinguished team of like-minded scientists, working on the conductive effects of X-rays on gases, the discovery of the electron, radioactivity and the identification of different types of radiation.

It was Rutherford who coined the terms 'alpha' and 'beta' to describe the mysterious particles emitted by radioactive material, and proved that radioactive material, no matter what its size, always took the same length of time for half the sample to decay to a neutral state. He called this feature radiation's 'half-life'. Nuclear physics and the link between energy and matter were converging rapidly, especially after a young German scientist called Albert Einstein published a theoretical treatise in 1905 linking mass and energy in the formula $E = mc^2$. Not to be outdone, in 1913 the brilliant Dane Niels Bohr published a model of what really goes on inside the atom. Bohr showed that atoms were not actually solid balls but nothing less than mini solar systems of whirling particles. The result of all this activity was that, by the 1920s, nuclear physics and the study of the atom were making dramatic technical advances.

By 1932, James Chadwick had revealed the neutron inside the atom, and in 1934 Enrico Fermi demonstrated that these neutrons could be used to bombard and split many different types of atoms and, bafflingly, yield results lighter than their original elements. What had happened to the missing parts of the atom? The discoveries raised as many questions about the structure of matter as they provided, inspiring further research. Not to be outdone, the Germans made a startling discovery in 1938 when Otto Hahn and Fritz Strassman, replicating an experiment that Fermi had carried out in 1913, proved that when the atoms are split, energy is released. That confirmed Einstein's theory, put forward 33 years earlier: energy and matter were two sides of the same coin. By 1939, atomic scientists everywhere were agreed that splitting atoms released unimaginable amounts of energy. Pandora's box was now well and truly open. The question, as war broke out in autumn 1939, was what to do with this knowledge.

War, not necessity, is the real mother of invention. Right from the start, however, the Führer Adolf Hitler's key nuclear experts had begun to disappear from Germany. Many of the country's top physicists had already voted with their feet. Jewish scientists realised that the Nazis were, at best, sidelining them and, at worst, rounding them up for despatch to camps as 'sub-human undesirables'. The smart ones emigrated, mostly to the United States. The 1930s émigrés included Einstein, Theodore von Kármán, John von Neumann, Leo Szilard and Edward Teller. Even the Italian Enrico Fermi, discoverer of induced radioactivity by neutron bombardment and winner of the 1938 Nobel Prize in Physics, fled. He had dared to marry a Jewish woman and feared anti-Semitic reprisals from Italy's fascist leader, Benito Mussolini. These eminent scientists continued their atomic research and discussions overseas. The United States and Britain were the beneficiaries.

The British were first to spot the real possibilities, and dangers, of an atomic weapon. In true Whitehall fashion, a committee was founded (named 'MAUD' after Niels Bohr's housekeeper) in response to a memorandum of March 1940 by Rudolf Peierls and Otto Robert Frisch, two refugee physicists working at the University of Birmingham. They warned that a small sphere of pure uranium-235 could have the explosive power of thousands of tonnes of TNT. In response to the MAUD Committee report on this interesting conclusion, a British nuclear weapons programme was launched with a specialist new directorate; for security purposes, it was given the deliberately misleading name of 'Tube Alloys'. The hunt for an atom bomb was on.

Across the Atlantic, by the summer of 1939, six months after the discovery of uranium fission, there was still no official US atomic energy project. Most American physicists doubted that atomic energy or atomic bombs were scientific possibilities, despite the fact that American newspapers, magazines and even comics were openly discussing a future bristling with the prospect of atomic energy.

Such governmental complacency received a sharp jolt in October 1939 when President Franklin D Roosevelt received a letter signed by Einstein – composed with considerable input from the latter's colleague, the Hungarian-born physicist Leo Szilard – that had been written the previous August. Szilard had been warning of the dangers of an atomic weapon for several years. He had noted that Nazi Germany had stopped the sale of uranium ore from occupied Czechoslovakia and key German scientists were known to be experimenting with nuclear fission. Uranium and nuclear fission? To scientists in the know, the danger was obvious. Szilard's concern was that Hitler could gain an unbeatable lead in developing some kind of atom bomb. Now, he needed Einstein's help.

Initially, he asked if Einstein would warn the Belgians to stop the large stockpile of pitchblende and uranium ore in the Belgian Congo from falling into Nazi hands. Einstein, who was willing to write to President Roosevelt, agreed and signed what was to become the key letter in the story of the atom and America's attempt to build an atomic bomb. The letter was a warning about the dangers of Germany acquiring massive destructive power and the need for the USA to investigate nuclear weapons. This was a tough letter for Einstein to put his name to, for he was a lifelong pacifist, vehemently opposed to scientists inventing weapons. But even Einstein now realised that pacifism and appeasement would not succeed against Hitler's Nazis, who viewed violence as the key to conquest. The letter is, therefore, one of the most important documents in history, because it encouraged the American president to start thinking about building an atomic bomb.

```
                                        Albert Einstein
                                        Old Grove Rd.
                                        Nassau Point
                                        Peconic, Long Island

                                        August 2nd, 1939
F.D. Roosevelt,
President of the United States
White House
Washington, D.C.



Sir:

        Some recent work by E. Fermi and L. Szilard, which has been com-
municated to me in manuscript, leads me to expect that the element uran-
ium many be turned into a new and important source of energy in the im-
mediate future. Certain aspects of the situation which has arisen seem
to call for watchfulness and, if necessary, quick action on the part
of the Administration. I believe therefore that it is my duty to bring
to your attention the following facts and recommendations:
        In the course of the last four months it has been made probable -
through the work of Joliot in France as well as Fermi and Szilard in
America - that it may become possible to set up a nuclear chain reaction
in a large mass of uranium, by which vast amounts of power and large quant-
ities of new radium-like elements would be generated. Now it appears
almost certain that this could be achieved in the immediate future.
        This new phenomenon would also lead to the construction of bombs,
and it is conceivable - though much less certain - that extremely power-
ful bombs of a new type may thus be constructed. A single bomb of this
type, carried by boat and exploded in a port, might very well destroy
the whole port together with some of the surrounding territory. However,
such bombs might very well prove to be too heavy for transportation by
air.
```

-2-

The United States has only very poor ores of uranium in moderate quantities. There is some good ore in Canada and the former Czechoslovakia, while the most important source of uranium is Belgian Congo.

In view of this situation you may think it desirable to have some permanent contact maintained between the Administration and the group of physicists working on chain reactions in America. One possible way of achieving this might be for you to entrust with this task a person who has your confidence and who could perhaps serve in an inofficial capacity. His task might comprise the following:

    a) to approach Government Departments, keep them informed of the further development, and put forward recommendations for Government action, giving particular attention to the problem of securing a supply of uranium ore for the United States;

    b) to speed up the experimental work, which is at present being carried on within the limits of the budgets of University laboratories, by providing funds, if such funds be required, through his contacts with private persons who are willing to make contributions for this cause, and perhaps also by obtaining the co-operation of industrial laboratories which have the necessary equipment.

I understand that Germany has actually stopped the sale of uranium from the Czechoslovakian mines which she has taken over. That she should have taken such early action might be understood on the ground that the son of the German Under-Secretary of State, von Weizsäcker, is attached to the Kaiser-Wilhelm-Institut in Berlin where some of the American work on uranium is now being repeated.

                                                       Yours very truly,

                                                       (Albert Einstein)

Delayed by the crisis created by the Nazi invasion of Poland, the letter reached the President only two months later, delivered by hand on 11 October 1939, six weeks after the outbreak of World War II. Washington reacted strongly to this alarm call, and the following months saw a major effort by Roosevelt's administration to begin investigating the problem of developing an atomic weapon and the setting up of nuclear committees in key US government departments. Newly funded research and inter-agency coordination of the results smoothed the path. In 1941, the US National Defense Research Council opened a liaison office with Tube Alloys and the British government to exchange findings. Developments received an even bigger boost when the Japanese bombed Pearl Harbor on 7 December 1941 and began its invasion of southeast Asia. Suddenly the development of an atomic bomb became a key US national defence priority.

In January 1942, Roosevelt authorised funding for a full-scale, no-expense-spared development study into the feasibility of an atomic weapon. In September 1942, research on nuclear fission was made the primary responsibility of the United States Army Corps of Engineers' Manhattan District office. They directed an all-out atomic development programme, thereafter known as the 'Manhattan Project'. In one of history's great ironies, Einstein never worked on the enterprise for which he had personally lobbied the US President. The US security authorities had denied the great nuclear scientist a work clearance in July 1940, claiming that Einstein's avowed pacifist leanings and celebrity status made him a national security risk.

The task of managing what was clearly a vast project was given to Colonel Leslie R Groves, of the US Corps of Engineers. Groves, a no-nonsense West Point graduate, was promoted to the rank of general and told to get on with it. One of his first major decisions was to select Oak Ridge, Tennessee, to be the production site for isotope separation. An isotope is an atom that has somehow acquired an extra neutron particle and therefore has different properties. For example, uranium-235 (also referred to as U-235) is highly radioactive: much more so than natural uranium (U-238). Isotopes were vital because uranium-238 doesn't go bang easily; but its isotope uranium-235 does, once its reluctant atoms are forced together to form a critical mass. The first task was to manufacture a useful quantity of the uranium-235 isotope as the active energy source.

An unexpected bonus from the isotope process was the discovery that a new radioactive material was created by neutron bombardment of uranium-238. This element was named plutonium and given the designation plutonium-239. Plutonium offered the best potential for an atomic weapon. It was easier to process and manufacture than uranium-235, had greater explosive power and, with its smaller size and weight, was more suited to being carried as a bomb by the aircraft of the day.

A key early decision was to consolidate atomic development activities in Chicago. Fermi and his scientific colleagues moved from New York, and began operation at the University of Chicago with a mandate to produce a nuclear chain reaction by the end of 1942. For raw materials, Fermi assembled a huge pile of 6 tons of uranium and 250 tons of graphite for use as a moderator. Manually operated cadmium rods were used to control any reaction.

This experimental 'pile', to prove the feasibility of a controlled chain reaction, was assembled in an abandoned squash court.

On 2 December 1942, with forty-two observers anxiously watching the instruments, the pile went critical, achieving a self-sustaining nuclear reaction for about four minutes until the control rods were slid back in to damp it down. The assembled scientists were delighted. A controlled atomic reaction really was possible.

All this atomic research cost a vast amount of money and sucked in huge resources. One of the least-known challenges was that there wasn't enough copper in America to build the massive magnets needed for the project's electromagnetic system. The problem was solved by ordering the bewildered US Treasury Depository in West Point, New York, to hand over several tons of solid silver as a conductor instead. By the end of 1943, the project seemed ready, but the all-important magnets were still found to be faulty and had to be completely rebuilt, causing another delay. By January 1944, the system was suitable for operation. This coincided with the decision to step up the production of plutonium and the discovery that thousands of diffusion tanks for the gaseous-diffusion process were required – causing yet more delay. In fact, it wasn't until early 1945 that Oak Ridge in Tennessee began shipping weapons-grade uranium-235 to Los Alamos in the desert of northern New Mexico, where weapon development was taking place.

By early 1945, the Manhattan Project had grown out of all recognition. Groves's atomic empire now stretched from New York to Hanford in Washington State, and from Chicago to the heart of the operation, Los Alamos. J Robert Oppenheimer was the director of this remote desert site perched on a rugged mesa about fifty-five kilometres northwest of Santa Fe. Los Alamos became a truly international effort, certainly on the scientific side. Experts from all over Europe worked there and the British effectively combined their efforts completely into the American project.

At the Quebec Conference in August 1943, Roosevelt and British Prime Minister Winston Churchill had signed the Quebec Agreement, which merged their two national projects. The document established the Combined Policy Committee to coordinate the two nations' efforts. The Hyde Park Agreement of September 1944 further extended this cooperation into the post-war period. With the sharing of intelligence, nuclear cooperation was to form the bedrock of the 'special relationship' (the phrase was coined by Churchill) between the USA and UK – one based not on politics, as is often assumed by lazy journalists, but on sharing war-winning secrets.

While this huge effort had been going on in the USA, other warring nations were chasing their own dreams of an atomic weapon. In Germany, atomic physics and its potential as a weapon was as well understood as it was in Britain and the United States. German scientists, however, were wrestling with two intractable problems: lack of interest from the Nazi leaders and lack of funding and resources. In June 1942, the Nazi engineer supremo Albert Speer discussed with Hitler the possibility of developing an atomic bomb, but came away with no clear conclusions or funds. Hitler only wanted short-term gains, not long-term plans. German atomic research would continue to fall behind.

The Germans did hold one trump card – they had a lot of 'heavy water' (D_2O). As early as October 1939, Soviet physicists concluded that this and carbon were the best moderators to control any uranium reactor, and in August 1940 they submitted a plan to the Russian Academy of Sciences calculating that a massive 15 tons of heavy water were needed for a working reactor. Other scientists, working in parallel to this research, confirmed that heavy water was crucial to control any nuclear reaction, because it slows down neutrons and gives them time to react with the refined weapons-grade uranium-235, rather than with natural uranium-238, which captures neutrons without reacting and fissioning. Heavy water was essential for the development of nuclear reactions.

The Allies turned their concerns to a large stock of heavy water produced and stored at Vemork in Norway. Several attempts by British special forces groups to sabotage the plant had failed, but it was finally bombed and destroyed by the United States Army Air Forces in November 1943. The Nazis then decided to move the precious heavy water to safety in Germany. The British and Norwegians promptly sabotaged the ferry, sending its cargo to the bottom of a Scandinavian lake. After the war, a member of German Army Ordnance confirmed that the loss of heavy-water production in Norway was the main factor in Germany's failure to achieve a self-sustaining atomic reactor. That 1944 attack effectively ended any German ambitions for an atom bomb.

On the other side of the world, in Japan, the military was in charge of atomic-weapon research. The Imperial Japanese Army had been following international scientific literature and, in 1938 and 1939, noted the discovery of nuclear fission. In

October 1940, a Japanese General Staff report concluded that Japan had access to sufficient uranium in Korea and Burma to make an atomic bomb. Japan's wartime A-bomb efforts were headed by Yoshio Nishina, who had worked in Copenhagen with atomic pioneer Niels Bohr. Nishina's team had the necessary expertise and, in April 1941, the Imperial Army Air Force authorised research to develop an atomic bomb, but it came to nothing. Japanese nuclear research and development were severely handicapped by the effects of the war on the country's industrial economy. By late 1944, most Japanese scientists realised that they would never be able to build a bomb in time to affect the outcome of the war. The project withered and died, despite discredited post-war rumours that Tokyo had developed a workable bomb in Manchuria.

The USSR, on the other hand, had realised the importance of atomic weapons even before the war and pumped resources into a major scientific programme to develop a weapon. By 1939, they had begun work on creating an atomic bomb. Igor Kurchatov, a senior Soviet physicist, had alerted the party hierarchy to the military potential of nuclear fission. The German invasion in June 1941 temporarily halted any nuclear work, disrupting research at key institutes as scientists fled east of the Urals to avoid the invading Nazis. Work on a Soviet atomic weapons programme only resumed in early 1943. Soviet efforts were greatly helped by the intelligence resulting from penetration of Britain and the USA by the NKVD (the forerunner of the KGB), particularly via a network of well-placed spies in the British establishment.

By 1944, the scientists at Los Alamos had worked out that there were two possible ways to make a nuclear explosion. The key problem was 'critical mass'. Uranium, even in its artificial U-235 state, would not react unless forced together for long enough to overcome its natural tendency to fly apart. It then had to be forced to stick together for long enough to start a chain reaction, albeit that the time span is one of only milliseconds. The scientists discovered that the answer was to force two sub-critical halves together and then fire a neutron source to begin a reaction as they collided. Originally this was done via a 'gun system', whereby the two halves of the nuclear core were fired towards each other by explosive charges to collide in a critical mass. Scientists also discovered another method, using a ball (or 'core') of fissionable uranium surrounded by a number of small explosive charges that simultaneously blew inwards, squeezing the ball into a critical mass, which then spontaneously detonated, with a little encouragement from a neutron initiator. This second approach was called the 'implosion method'. The basic gun-type system was considered the more reliable and was mechanically simpler, as the implosion method required simultaneous detonation of the lenses around a perfect sphere and was technically more risky. But by early 1945, an atom bomb was theoretically possible.

The final problem was how best to deliver such a bomb. American technical expertise, coupled with unlimited resources and a characteristic 'can-do' attitude, had the solution to hand. The logical choice was the new Boeing B-29 Superfortress bomber with its 4,830-kilometre (3,000-mile) range, up to 32,000-feet (9,750-metre) high-altitude performance, and the ability to carry bulky and heavy bomb loads. In late 1944, a test flight programme confirmed the B-29's suitability to become the first atom bomber.

A new composite bomber group of B-29s, the USAAF's 509th, was formed under the command of Lieutenant-Colonel (later Colonel) Paul W Tibbets. As a veteran of B-17 missions over Europe, Tibbets was highly qualified for the position. In October, the 509th began receiving its new B-29s and commenced an intensive training schedule designed specifically to prepare for a high-altitude release of the bomb. It also practised various escape manoeuvres to avoid any shock effects from the unknown blast wave from the new bomb. Finally, in May 1945, just as the European war collapsed with the defeat of Germany, the 509th deployed to the island of Tinian in the Marianas, 2,400 kilometres (1,500 miles) south of Japan. All thoughts now turned to the Japanese war and how to end it as quickly as possible.

The USA was ready to do just that. At 5:29 am on 16 July 1945, the United States Army, as part of the Manhattan Project, detonated a test device atop a thirty-metre tower in the desert some fifty-five kilometres southeast of Socorro, New Mexico, now part of White Sands Missile Range. Many of the watching scientists were sceptical. Few had any idea just how big – or small – the explosion would be, or even it there would be an explosion at all.

An impromptu syndicate among the watching scientists bet that a nuclear 'fizzle' was the most likely outcome. Others disagreed: Oppenheimer, the director at Los Alamos, chose a small explosion of only 0.3 kilotons of TNT as being the most likely. Enrico Fermi offered to take wagers on whether the Earth's atmosphere would ignite and, if so, whether the chain reaction would destroy the assembled throng, or just the state, or perhaps even start a chain reaction that would incinerate the entire planet, much to the unease of his colleagues.

All except an electromagnetic expert called Isodor Rabi lost their stake. He arrived late and found the only sweepstake entry left was for an 18-kiloton explosion. When the bomb was detonated, the blast was rated at 18.6 kilotons, and lucky Rabi scooped the pot. The device itself, code-named 'Trinity', demonstrated the awesome power of the atom. The roar of the shock wave took 40 seconds to reach the observers. It was felt more than 160 kilometres (100 miles) away and the mushroom cloud reached 12 kilometres (40,000 feet) high. Oppenheimer later recalled that:

We knew the world would not be the same. A few people laughed, a few people cried. Most people were silent.

I remembered the line from the Hindu scripture, the *Bhagavad Gita*: Vishnu . . . takes on his multi-armed form and says, 'Now I am become Death, the destroyer of worlds.'

I suppose we all thought that, one way or another.

On 6 August 1945, Colonel Paul Tibbets's B-29, *Enola Gay*, named after his mother, dropped a single bomb, code-named 'Little Boy', containing 64 kilograms of uranium-235 on the Japanese city of Hiroshima. It took 45 seconds to fall from the aircraft, flying at about 31,000 feet (9,450 metres), to its detonation height of about 1,900 feet (580 metres). The 16-kiloton airburst killed an estimated 100,000 Japanese soldiers and civilians, with thousands more deaths to follow.

On 9 August 1945, a B-29 named *Bockscar*, piloted by Major Charles W Sweeney, dropped a second atom bomb – 'Fat Man', with a yield of about 21 kilotons – on the city of Nagasaki, killing an estimated 75,000 people.

The world had changed for ever. The age of the atom had arrived.

CHAPTER 2

RUTHLESS RAYS: RADIATION – SILENT BUT DEADLY

'When one studies strongly radioactive substances, special precautions must be taken [. . .] Dust, the air of the room, and one's clothes, all become radioactive.'
– **MARIE CURIE, PHYSICIST AND CHEMIST**

Assuming that you haven't been vaporised, blown to bits, wounded or disfigured by the blast, or fried half to death by the heat pulse, nuclear radiation is the real killer. It is this that really sets atomic weapons apart from conventional explosions and makes them so lethal. It is radiation that makes nuclear weapons – and nuclear accidents – a mortal threat. Silent, invisible nuclear rays spread out to attack the cells that make up all living things, killing, damaging organs and tissue, causing disease, and even altering the molecular structures to change or mutate living cells into new, unwelcome life forms. Ukrainian biologists report some very strange and twisted animals now roaming the restricted zone around Chernobyl.

The substances that give out these deadly rays are termed radioactive, and they have the ability to contaminate everything with which they come into contact, live or inert. Soil can become irradiated enough to glow in the dark from the deadly energy stored in radioactive contamination. Nuclear radiation is the deadliest threat of all; it cannot be seen but can linger for centuries.

Plutonium for example, decays very slowly indeed. At Chernobyl in the Ukraine, 'hot' particles of plutonium-241 scattered across the landscape are actually decaying into an even more toxic isotope, americium-241. This, in turn, decays into neptunium-237, another energetic alpha emitter and one with a half-life of more than two million years. The contamination area around ground zero is estimated to make the area immediately surrounding the power plant itself uninhabitable for a staggering 20,000 years. Even in the wider exclusion zone, with a radius of thirty kilometres, the irradiated soil contaminated by deposits of strontium-90, caesium-137 and plutonium will keep the area off limits for up to three thousand years, according to Ukrainian experts. The implications and long-term dangers of nuclear radiation are mind-boggling.

The full horror of radiation's effects became apparent immediately after the Hiroshima bomb was dropped. On 24 August 1945, just eighteen days after 'Little Boy' devastated the city, a famous Japanese actress called Midori Naka, who had been some 650 metres from ground zero at Hiroshima, died in Tokyo. By some miracle she had survived the bomb blast.

> I was in the kitchen... When a sudden white light filled the room, my first reaction was that the hot water boiler must have exploded. I immediately lost consciousness. When I came to, I was in darkness and I gradually became aware that I was pinned beneath the ruins of the house. When I tried to work my way free, I realised that apart from my small panties, I was entirely naked. I ran my hand over my face and back: I was uninjured! Only my hands and legs were slightly scratched. I ran just as I was to the river, where everything was in flames.

> I jumped into the water and floated downstream. After a few hundred yards, some soldiers fished me out.

A few days later, she fled to Tokyo and on 16 August, Naka consulted Dr Matsuo Tsuzuki, at the time the foremost radiation expert in Japan, at the Tokyo University hospital. He diagnosed radiation sickness and prescribed repeated blood transfusions to save her life. Naka's hair began to fall out and her white blood cell count sank from the normal count of 8,000 to 300–400 (some sources suggest the figure was between 500 and 600 white blood cells), much to the surprise of the doctors. Her red blood cell count was at the 3,000,000 level, around 60 per cent of normal levels. By 21 August, her temperature had risen to 41°C, around five degrees above normal, and her pulse rate shot up to 158; a normal adult pulse rate is between 60 and 100 beats per minute. On 23 August, some about thirteen livid purple patches appeared on her body. She died the next day.

Midori Naka was the first person in the world whose death was officially recorded as 'atomic bomb disease' or as we now know it, radiation sickness. The silent killer, nuclear radiation, had claimed yet another victim, yet scientists were not surprised. The medical community had known about radioactivity's dangers for years

In December 1895, working in his lab in the University of Würzberg, Wilhelm Röntgen had noticed invisible rays that fogged photographic plates and seemed to be able to travel through solid wood or flesh. A few months later in France, Henri Becquerel discovered that minerals containing uranium also gave off rays. Röntgen's X-rays amazed scientists; Becquerel's rays, which were considered to be just a weaker version of Röntgen's, were pushed on to the back burner by researchers. To this day the prime measurement of radiation is the 'röntgen' (also 'roentgen'). This is sometimes expressed as 'rads': one röntgen of external radiation exposure is often described as one rad of radioactivity absorbed by material or tissue. The other commonly used measurement for radiation is the 'röntgen equivalent man' (abbreviated to 'rem').

The acute effects of radiation exposure were first seen in 1896 when Nikola Tesla, inventor of the AC electrical current and the dynamo, among many other things, purposely subjected his fingers to X-rays. He reported that this caused burns to develop, although he did not link this to radiation. These early hit-or-miss discoveries drew nuclear pioneer Marie Curie to her life's work. Working in Paris, she decided to probe deeper into the mysteries of radioactivity, investigating the strange uranium rays. That curiosity would one day be her undoing. With little scientific literature to read at the time, she began to experiment with some very dangerous substances.

Marie noticed that pitchblende, the black ore from which uranium is extracted, emitted powerful rays, stronger than uranium. Intrigued, she deduced that the very large readings she was getting could not be caused by uranium alone – there must something more radioactive hiding in the pitchblende. Since nobody had ever found it before, it could only be present in tiny quantities.

Along with her husband, Pierre, she began researching this unknown source of radiation. They laboriously ground up samples of pitchblende, dissolved them in acid and, using Bunsen burners and the standard school lab chemistry techniques of the day, began to separate the different elements in the pitchblende. They finally ended up with small pile of black powder, 330 times more radioactive than uranium. Marie Curie was Polish by birth, so she christened this new chemical element, polonium, atomic number 84.

More intriguing still, the Curies found that there was some extremely radioactive liquid left behind after they had extracted the polonium. They realised that this was yet another new element, far more radioactive than uranium or polonium, and present only in tiny quantities. In 1898, the Curies published their evidence of a new element, which they christened 'radium'. There was only one problem: no one had actually seen or proved that their great discovery existed. The scientific world remained unimpressed.

Marie then located a factory in Austria that specialised in removing uranium from pitchblende for industrial use and bought several tons of the seemingly worthless waste product. Pitchblende waste turned out to be even more radioactive than the original pitchblende, and was much cheaper. Marie set about processing the waste by hand to extract the tiny quantities of radium. This involved heavy and physically demanding work on an industrial scale, with 20-kilogram sacks of filthy black waste that had to be ground to a powder, then dissolved in a chemical reagent, filtered, precipitated, then re-dissolved and turned into new crystals.

The resulting amounts were smaller and smaller but eventually the Curies had the first-ever sample of pure radium. By 1902, just 0.1 grams of pure radium was refined from several tonnes of pitchblende waste, and by 1910 Marie Curie and French chemist André-Louis Debierne had isolated the metal itself. It was lethal stuff. Radium is intensely radioactive. Although the Curies' labours won them the 1903 Nobel Prize in Physics – jointly with Henri Becquerel – for their work on radioactivity, it had had exposed them to serious health risks. Both of them complained of nausea and exhaustion. Although they didn't know it, they were suffering from radiation sickness.

Marie Curie's work would eventually kill her. On 4 July 1934, she died of aplastic pernicious anaemia, a condition she developed after years of exposure to radiation. Like many idealistic scientists, she saw only the possibility of scientific progress, never the potential dangers of her discoveries, and is recorded as saying, 'I am one of those who think, like Nobel, that humanity will draw more good than evil from new discoveries.'

Her legacy lives on as the 'mother of radium'. It would turn out to be a murderous and dangerous discovery.

By the outbreak of World War I, hard-eyed businessmen had recognised the commercial benefits of radium. Outside the closed world of nuclear physics, radium wasn't thought to be particularly dangerous, and there were exciting new products to be marketed, including radium bath salts, radium candy and even radium cigarettes. And radium also had a unique and useful tendency to glow in the dark – what better material to paint on the hands of watches? By the early 1930s, small armies of young women were employed in factories painting fine watch hands with radium compounds.

No one absorbed more of this toxic material than the 'radium girls'. They regularly licked their brushes to ensure they had a fine point, thereby ingesting dangerous quantities of radioactive radium into their bodies. The radium then disobligingly settled in their bone marrow. There, the otherwise weak alpha radiation could interfere directly with the body's production of red corpuscles. Many of the workers developed anaemia and bone cancer. By 1927, a group of radium girls had banded together to sue the US Radium Corporation, and were awarded an out-of-court settlement; it was not until 1938 that another group successfully sued their employer in court. Some safety measures were thereafter taken, but radium continued to be used on items, for instance, watch hands, until the 1960s.

We now understand the dangers of nuclear radiation far better than the unfortunate radium girls. The horrors of Hiroshima, Nagasaki, Chernobyl and Fukushima have given us rather more examples of the dangers of radiation than we really need to grasp the lessons. We now know that all three basic atomic particles that are radiated are dangerous in their different ways.

The weakest by far are alpha rays. Alpha rays are actually high-speed particles. They are effectively an atomic nucleus stripped out of its home by a nuclear event (like splitting the atom or an atom bomb) and sent whizzing off into micro space. Although alpha radiation travels very fast, it can easily be blocked or shielded because, as the particles move through matter, they are constantly attracted to interacting and joining with other charged particles, and swiftly lose their energy. A good analogy is the cue ball striking the fifteen reds on a snooker table. After that, it doesn't go far.

The result is that even air can slow down or stop alpha particles and a single sheet of paper or skin stops them dead. Alpha rays are not, therefore, hazardous externally. But – and it is a seriously alarming 'but' – if alpha particles are ingested, or inhaled, then the highly toxic radioactive material nestles against living tissue deep inside the body and can wreak havoc. That is exactly what happened to the unfortunate Russian defector Alexander Litvinenko in London in 2006. He had unwittingly drunk a cup of tea spiked with polonium-210, generously supplied by Russia's SVR or GRU intelligence agencies. Their mission accomplished, his assassins fled back to Moscow, leaving Litvinenko to his painful, lingering death.

Even worse are free-ranging beta particles, or rays. They have a medium penetrating power and can be blocked by a sheet of light metal, such as aluminium or even plastic. Beta rays are much lighter than alpha radiation, being high-energy particles spun off from a split atom at a measured speed of 270,000 kilometres per second; the beta particle is effectively an orphan atomic electron looking for a new home. They are more dangerous than alpha radiation because they have more penetrative power, thrusting far deeper into living matter and capable of irradiating soil or the fabric of buildings. In living creatures, beta radiation damages the chemical links between the molecules when it hits a living cell, and can even cause some permanent genetic change in the cell nucleus. If that damage occurs within the reproductive organs of animals or humans, then the damage can be passed on to future generations in the form of birth defects or mutations.

Of the trinity of types of unholy radiation, the most dangerous are the very high-energy gamma rays. These have no mass and no charge. They are not particles and are more akin to very powerful X-rays. They belong in the electromagnetic spectrum alongside light and true X-rays as photons or light waves, but are much more energetic and dangerous. Gamma rays are produced by the hottest and most active objects in the universe, such as neutron stars and pulsars, supernova explosions and the mysterious regions around black holes. They can be lethal, as they slow down on contact with other matter, damaging living cells and transferring their energy to surrounding cell components or anything else in their path. This makes gamma rays potentially deadly to living material. Their extremely high energy allows them to penetrate just about anything, from wood and metal to bone. They can destroy living cells, produce gene mutations and cause cancer. Only a dense material, such as lead shielding, can block them and stop them in their tracks.

Like beta rays, gamma radiation can also cause 'induced radiation'. This is a serious and very dangerous hazard associated with nuclear explosions. In summary, some nuclear radiation can cause other things to become radioactive. Radiation, travelling as a particle (alpha, beta) or electromagnetic wave (gamma), attacks the atomic particles it hits and makes them radioactive in turn by interfering with their atomic structure. It 'ionises' them; irradiation from the different forms of ray is known as ionising radiation.

In 1934, Marie Curie and her daughter Irène discovered that if they bombarded a stable material, the lighter elements continued to emit radiation long after the radiation source was removed. Radiation, it seemed, induced radiation in other materials. This was graphically confirmed at the White Sands Trinity test of July 1945, when the surface around ground zero was fused into a glass-like slag that was dangerously radioactive. It confirmed that radiation could breed or induce radiation in other things. Given the half-life of nuclear material, it could stay radioactive for a long time before breaking down far enough to become neutral and safe.

Understanding this induced half-life is vital, because it explains why nuclear radiation is so dangerous to life for so long. Soil, buildings and anything that has been exposed to and irradiated by a blast of nuclear radiation can harbour radioactivity for decades, the newly radioactive atoms like invisible death rays, waiting to pounce on the unwary. Studies of populations that have experienced unusually high levels of radiation, such as those brought about by the 1945 atomic bombings of Hiroshima and Nagasaki or the 1986 Chernobyl disaster, record thousands of people suffering radiation sickness, and especially cancer, years later.

In 1970, Dr Alice Stewart, director of the Nuffield Institute of Social Medicine, was one of the first experts to alert the world to the dangers of this persistent low-level radiation. The atomic age had been born in secrecy, and for two decades after Hiroshima, scientists and governments, under the all-embracing cloak of state security, had concealed vital information about the risks to human health posed by radiation. Stewart concluded that low-level radiation from fallout and nuclear-waste disposal, especially among nuclear workers and the public, was more dangerous than had been publicly admitted.

Concentrating her research on radiation damage to genes at the molecular level, Stewart infuriated the medical and scientific establishment by warning that she suspected the long-term dangers of nuclear radiation were much worse than they had acknowledged. The professional opposition to this turbulent scientist was immediate, personal and well-organised.

Worse was to come. In 1974, she was invited to the United States by Professor T F Mancuso to become a consultant on a major US government study into the health of nuclear workers at Hanford, which had produced plutonium for the Manhattan Project. Designed to parallel a study into the long-term health effects of low-level radiation on the survivors from Hiroshima and Nagasaki, this research was vital to the safety of workers in the nuclear industry. The industry was regulated by federal law to work within the exposure levels and standards laid down by the US government-approved International Commission for Radiation Protection (ICRP). The Mancuso-Stewart radiation study was intended as a review and validation of those standards. Instead, it became an exposure of too-lax regulation by the government medics and nuclear scientists, revealing the true risk to nuclear workers.

To the horror of these powerful nuclear lobbies, within and outside government, the study concluded that the low-level radiation standards being imposed on nuclear workers and the public by fallout and nuclear-waste disposal had far more serious consequences than had been officially admitted. Its findings made for uncomfortable reading to Washington's medical and scientific administration. Dr Stewart's analysis revealed the long-term incidence of cancer was roughly ten times more than the official predictions from exposure to nuclear radiation, both at Hiroshima and at Hanford. The dread words 'lawsuits' and 'compensation' began to be bandied around.

Faced with the threat of prized contract dollars disappearing down the plughole, an outraged nuclear establishment promptly rubbished Stewart's findings and the US government sacked Professor Mancuso. The full survey results were never published in their original form and the use of outside consultants was promptly banned. We now know beyond doubt that Dr Stewart was correct – there is a clear link between ionising radiation and cancer. The chilling truth is that nuclear weapons and nuclear accidents undoubtedly pose dire long-term threats to human and all other life on our planet. The fundamental facts are now brutally clear:

- Whether from nuclear power generation or an atomic bomb, energy is generated from atomic fission caused by the collision of neutrons in uranium or plutonium.
- When uranium-235 undergoes nuclear fission, it produces new lethal radioactive substances such as iodine-131, caesium-137, strontium-90 and cobalt-60.
- Release of such radioactive substances into the environment in large quantities causes serious health effects from radiation exposure.
- The harmful effects caused by the longevity or half-life of these nuclear substances can last a very long time.
- Iodine-131 has a physical half-life of eight days, meaning that it takes about eight days for 100 radioactive iodine-131 nuclei to be reduced to half that number (50 nuclei). Iodine binds on and affects the thyroid.
- The close relationship between exposure to radioactive iodine and increased risk of thyroid cancer has been confirmed as a result of the Chernobyl nuclear power plant accident.

Other nuclear by-products are just as deadly:

- Caesium-137 has a half-life of thirty years. It is readily absorbed into muscle tissue.
- Strontium-90 has a half-life of twenty-eight years. It is readily absorbed into bone tissue.
- Cobalt-60 has a half-life of five years.

The risks to the human body from these and other forms of radiation are only too real. Radiation is the lurking danger at the heart of any release of nuclear energy, from whatever source.

CHAPTER 3

OOPS! SOME EARLY ACCIDENTS

'The unleashed power of the atom has changed everything save our modes of thinking, and we thus drift toward unparalleled catastrophe.'
– ALBERT EINSTEIN

The dangers of nuclear science were demonstrated long before the first atomic test in 1945. Atoms could kill – one way or another. Those diligent researchers, the Curies, were perhaps its earliest victims. The whole family suffered as a result of their close contact with the world of the atom and radioactivity. First to go was Pierre Curie. He had been educated at home by his father and demonstrated a precocious talent for mathematics and science, matriculating at the age of sixteen to be taken on as a laboratory assistant at the Sorbonne. There he began his ground-breaking research on the formation of crystals. In 1894, the brilliant young physicist met an equally brilliant young Polish woman, Marie Skłodowska, and their marriage in 1895 marked the beginning of a remarkable scientific collaboration.

Between them, the Curies discovered polonium and radium and explored the phenomenon of radioactivity. Pierre concentrated on the physical study of the new radiation, proving the existence of particles electrically positive, negative and neutral, which Ernest Rutherford later identified as alpha, beta and gamma rays. Pierre was appointed a professor at the Sorbonne and in 1903 he and Marie shared the Nobel Prize in Physics.

On 19 April 1906, the absent-minded Pierre Curie, hurrying to cross a street in Paris, looked the wrong way, stepped into the path of the iron-clad wheels of a horse-drawn dray. It ran over his head, causing a fatal skull fracture. He was forty-six years old. Witnesses said he wasn't paying attention and had 'his head in the clouds'. According to Pierre Clerc, the Sorbonne lab assistant who identified Curie's body, 'He wasn't careful enough when he was walking in the street, or when he rode his bicycle. He was always thinking of other things.' When Pierre's father learned that his son had been killed crossing a Paris street in traffic on a rainy day, his first reaction was, 'What was he dreaming about this time?'

Nuclear physics had lost, in a brief, pedestrian in every sense, tragedy, one of its greatest scholars.

Pierre's unexpected death came as a bitter blow to Marie, but it was also a decisive turning point in her career. She threw herself into completing the scientific work that they had undertaken. 'Crushed by the blow, I did not feel able to face the future. I could not forget, however, what my husband used sometimes to say, that, even deprived of him, I ought to continue my work.'

She was appointed to the vacant professorship left on her husband's death, becoming the first woman ever to teach at the Sorbonne. In 1911, she was awarded the Nobel Prize in Chemistry and oversaw the establishment of the laboratories of the Institut du Radium at the University of Paris.

Despite these laurels, the hazards of nuclear radiation would claim her as another early victim. Long exposure to radiation eventually took its toll. As the years went on, Marie became afflicted with a number of mysterious ailments and medical problems. Today we know that exposure to radiation can cause damage to the eyes as the lenses become clouded, fogging over just like a photographic plate. By 1920, Marie learned that she had a double cataract. Her vision became so impaired that she had to write her lecture notes in huge letters and have her daughters guide her around. Only after four operations was she able again to carry out exacting lab procedures and drive a car. Marie herself began to suspect that her scientific curiosity was the source of her troubles, writing to her sister Bronya: 'Perhaps radium has something to do with these troubles, but it cannot be affirmed with certainty.'

The problem was that both the Curies, like other pioneering researchers and industrialists of the day, were unaware of the dangers of the material they were handling and the health effects of exposure to radioactivity. Pierre even carried a sample of radium in his pocket to demonstrate its properties to the curious. His wife carried samples of radium in her overalls and even kept a glass jar at home as a nightlight next to her bed. Both developed cracked and scarred fingers and became progressively sick with a number of unexplained ailments. Knowledge of the dangers dawned slowly: in 1925, Marie was a member of a French Academy of Medicine commission that recommended the use of lead screens and periodic tests of the blood cells of workers in industrial labs where radioactive materials were prepared.

Sadly for her it was too little, too late. Over the next few years, her health deteriorated; on some days, she wasn't able to work. In the spring of 1934, she paid a final visit to see her brother-in-law, Jacques, accompanied by Bronya. By May she had become chronically ill, so much so that one afternoon she was forced to go home from her lab. It was the last time she saw it.

The medical specialists remained baffled. Eventually, they diagnosed a blood disorder for which there was no cure. In the little time left to her, she worked on the manuscript of her book *Radioactivity*, which would be published posthumously in 1935.

Marie Curie died of radiation sickness on 4 July 1934, truly a victim of her own success. The director of the sanatorium near Mont Blanc, in which she spent her final days, observed: 'The disease was an aplastic pernicious anaemia of rapid, feverish development. The bone marrow did not react, probably because it had been injured by a long accumulation of radiations.'

In the event, she would be laid to rest twice. On 6 July 1934, she was interred in the cemetery near Versailles alongside her soulmate Pierre. More than sixty years later, their remains were removed and reburied in the Panthéon in Paris, the French national mausoleum. As a pioneer in so many ways, it was entirely fitting that Marie Curie should be the first woman to be afforded that right.

Her laboratory notebooks are rightly regarded as national and scientific treasures. Unfortunately, they are still so highly radioactive that they are stored in lead-lined boxes at France's Bibliothèque Nationale in Paris. Some are on display behind thick glass. All visitors have to sign a liability waiver and wear protective gear, as the items are contaminated with radium-226, which has a half-life of about 1,600 years. Madame Curie stands as a giant among scientists but, like so many of those involved with nuclear matters, she unwittingly paid a heavy price for her pioneering discoveries.

If Curie's death was caused by lack of knowledge, the next known fatality brought about by radioactivity was an accident caused purely by human error. It long predates any nuclear device. The first recorded nuclear accident was in Germany as part of the Nazis' attempt to build a bomb. On 23 June 1942, shortly after the new Leipzig L-IV experimental atomic pile (nuclear reactor) in eastern Germany was opened, there was a catastrophic nuclear fire that claimed four lives. The L-IV was conducting an experiment to prove the theory that with five tonnes of heavy water and ten tonnes of metallic uranium, a nuclear pile could sustain a fission reaction and provide energy. It did, although not quite in the way that Werner Heisenberg and his colleague Robert Döpel, who were demonstrating Germany's first signs of neutron propagation, intended. By a supreme irony, Heinrich Himmler, head of the SS, called Heisenberg a 'white Jew' – his racist term for an Aryan who acted like a Jew – and thought he should be 'made to disappear' into a concentration camp. Instead, he was busy working on the Nazis' atomic programme.

As part of the checks, the L-IV device was examined for a potential leak of heavy water. Somehow, in the course of that inspection, air entered the reactor and ignited the uranium powder inside, which boiled the water jacket. The resulting steam pressure destroyed the reactor and the laboratory became littered with the powder, which led to a fire. The world's first nuclear accident, it put an end to the project.

In 1945, after the war, ten German scientists, including three Nobel laureates, all involved in the Nazis' nuclear research, were interned for six months at Farm Hall, an MI5 safe house near Cambridge. All the rooms contained hidden microphones. Their now declassified revelations are startling. The transcripts show vividly that Heisenberg and his colleagues were stunned when they heard that the USA had dropped an atomic bomb on Hiroshima, and that Heisenberg's post-war claims that he tried to keep his research programme free of military control, and avoid working on an atomic bomb after the shock of the fatal accident at Leipzig IV, were probably true.

The Germans, however, were not the only ones to suffer from nuclear accidents. On 21 August 1945, a scientist called Harry Daghlian dropped a tungsten carbide brick and died as a result. Daghlian was another brilliant young physicist. In 1944, as a promising PhD student, he joined Otto Frisch's group at the research and development branch of the Los Alamos Laboratory of the Manhattan Project in New Mexico.

The mishap occurred as Daghlian was trying to construct a neutron reflector. He was piling a set of 5-kilogram tungsten carbide bricks around a plutonium core, the highly unstable element of nuclear bombs, but realised that he was making it more unstable with every brick he placed around the core, and was watching the radiation level closely. As he brought the final brick over the assembly, neutron counters monitoring the radioactivity levels alerted him to the fact that once it was placed, the system would become supercritical.

Understandably unnerved, Daghlian accidentally dropped the brick – which fell into the pile. This caused the reaction to go critical spontaneously. Daghlian reacted by attempting to knock the brick away, to no avail. He then pulled the tungsten carbide pile apart by hand to stop the reaction.

Fortunately, and despite the rules, there was no one else in the laboratory except Daghlian at that fatal hour. On hearing his warning shout, the guard sitting near the laboratory, who was only there to watch for a fire, promptly ran away.

Daghlian was estimated to have received a dose of 510 rads of neutron radiation. Despite intensive medical care, he developed symptoms of severe radiation poisoning and died twenty-five days later, following a painful battle. Daghlian's family members only learned about the real cause of his death many years later. They never knew that he was a member of the Manhattan Project; the US Army told them only that Daghlian had died in an accident in the chemical laboratory. Daghlian's work was not forgotten, though. In 2000, his hometown of New London erected a memorial stone and flagpole in Calkins Park with a citation that read: 'Though not in uniform, he died in service to his country.'

Harry Daghlian was not to be the only victim of what came to be known as the 'demon core'. Nine months to the day after his death, on 21 May 1946, the core on which he had been experimenting was the subject of another experiment that went badly wrong. Canadian physicist Louis Slotin, Daghlian's friend and colleague, was demonstrating the core's characteristics to visiting scientists at a secret laboratory in a canyon around five kilometres from Los Alamos. A well-researched article by Alex Wellerstein, entitled 'The Demon Core and the Strange Death of Louis Slotin', was published by *The New Yorker* in May 2016. The prestigious magazine has commented on many nuclear hazards over the years, and its account of Slotin's death cannot be bettered.

> Slotin was showing his colleagues how to bring the exposed core of a nuclear weapon nearly to the point of criticality, a tricky operation known as 'tickling the dragon's tail'. The core, sitting by itself on a squat table, looked unremarkable – a hemisphere of dull metal with a nub of plutonium sticking out of its centre, the whole thing warm to the touch because of its radioactivity... At that time, Slotin was perhaps the world's foremost expert on handling dangerous quantities of plutonium. He had helped assemble the first atomic weapon by hand, barely a year earlier...
>
> Slotin's procedure was simple. He would lower a half-shell of beryllium, called the tamper, over the core, stopping just before it closed and sent the nuclear mass critical. The tamper would reflect back the neutrons that were beginning to shoot off the plutonium, jump-starting a weak and short-lived nuclear chain reaction, on which the physicists could then gather data. Slotin held the tamper in his left hand. In his right hand, he held a long screwdriver, which he planned to wedge between the two components, to keep them apart.
>
> As he began the slow and careful process of lowering the tamper, one of his colleagues, Raemer Schreiber, turned away to focus on other work. . . . But suddenly he heard a sound behind him: Slotin's screwdriver had slipped, and the tamper had dropped fully over the core. When Schreiber turned around, he saw a flash of blue

light and felt a wave of heat on his face. A week later, he wrote a report on the mishap:

> The blue flash was clearly visible in the room although it [the room] was well illuminated from the windows and possibly the overhead lights . . . The total duration of the flash could not have been more than a few tenths of a second. Slotin reacted very quickly in flipping the tamper piece off.

That momentary slip of a screwdriver during the manual assembly of a critical mass of plutonium caused a prompt critical reaction and cost Slotin his life. He died on 30 May from massive radiation poisoning, with an estimated dose of 1,000 rads. Seven observers, who received doses as high as 166 rads, survived, yet three died within a few decades from conditions believed to be radiation-related.

In these incidents, Daghlian and Slotin were working with the same 'demon core'. It was subsequently melted down and combined with existing weapons-grade material. The accidents highlighted the dangers of interfering with the innermost workings of the atom. Right from the start, nuclear scientists and engineers were well aware that, with nuclear, they were playing with fire.

But the temptation to get in on the action proved too strong for friends and enemies alike. By 1942, the US Navy, jealous as ever of its position and influence over congressional funding in the snake pit of Washington politics, quickly became aware that there was some great secret scheme afoot. They determined to muscle in on the army's mysterious Manhattan Project. The results were to cost the admirals dear. Although it was officially excluded from the wider Manhattan Project on grounds of national security, the US Navy began its own research on nuclear reactors for possible use in ships and submarines.

In 1939, the physicist Ross Gunn had suggested that uranium could be utilised for 'submarine submerged propulsion' and 'could enormously increase the range and military effectiveness of a submarine'. A completely separate US Navy programme to research nuclear possibilities was set up in 1942, based at the Naval Research Laboratory (NRL) in Washington, DC, under the navy's chief scientist, physicist Philip Abelson. In particular, the NRL was investigating the use of liquid thermal diffusion to enrich uranium. The system 'seemed to work well' according to the navy team. Unfortunately, the Manhattan Project scientists had already confirmed that electromagnetic separation and gaseous diffusion were more effective but, owing to the lack of communication between the army and the navy, NRL was not aware of this.

Before long, the NRL had constructed an 11-metre-tall thermal diffusion 'column', comprising three pipes, one inside the other. 'The apparatus was run continuously with no shut down or break down whatsoever,' Abelson later reflected. 'Indeed, so constant were the various temperatures and operating characteristics that practically no attention was required to ensure successful operation. Many days passed in which operating personnel did not touch any control device.'

The following year, Abelson requested that an even bigger 300-column pilot plant be built at the Naval Boiler and Turbine Laboratory at the Philadelphia Navy Yard, though only a third of the columns would be completed. Understandably confident in their new system, the navy pushed on with their single-service research.

In early September 1944, three volunteers undertook a straightforward but dangerous task. Chemical engineer Peter N Bragg Jr, a thermal diffusion project worker called Douglas P Meigs and Special Engineer Detachment (SED) physicist Arnold Kramish had to repair a blocked tube in the navy's liquid thermal diffusion pilot plant. Between two of the pipe's tubes was liquid uranium hexafluoride, while a third pipe contained high-pressure steam.

But when Bragg and Meigs began trying to clear the blocked tube, an explosion occurred, breaking the tube and disastrously mixing the liquid uranium hexafluoride with the steam. The resulting hydrofluoric acid burned and killed the two engineers; Arnold Kramish suffered burns so severe he was not expected to live, but was to make a remarkable recovery at a naval hospital in Philadelphia.

The navy quickly covered up the accident. The experimental atomic facility at the Philadelphia Navy Yard was way above top secret and it was important not to let anyone know that it had links to nuclear research. At the Manhattan Project, an alarmed General Groves ordered a news blackout and demanded explanations from the navy. A press release described the event only as an 'Explosion at Navy Yard' and the press were fobbed off with stories of a minor accident. The coroner himself was not told the truth, which only came to light many years afterwards. The US Navy's nuclear efforts – and its congressional budget – were discreetly incorporated into the Manhattan Project.

The US Navy later figured large in what must be considered an early nuclear accident, albeit indirectly, but it is no exaggeration to say that it was the atom bomb that started the causal chain of what was to become the worst naval disaster in American history. On 16 July 1945, the heavy cruiser *Indianapolis* sailed from San Francisco to the island of Tinian in the Marianas, shortly after receiving confirmation of the success of the Trinity atom test. 'Little Boy's' irreplaceable 'gun-type' mechanism and uranium were sent aboard the cruiser under conditions of great secrecy for urgent transportation to Tinian.

The warship was commanded by Captain Charles McVay III, and key Manhattan Project members James Nolan and Robert Furman accompanied the atomic bomb parts on board.

USS *Indianapolis* was actually carrying the components of the 'Little Boy' uranium bomb, soon to be dropped on Hiroshima. The parts for 'Fat Man', which would fall on Nagasaki, were transported by aircraft. The uranium contained in the canisters aboard the *Indianapolis* was about half the total US supply. None of the crew knew what was contained in the crate and canister that were brought on board. The two men from the Manhattan Project who accompanied the components were not very convincingly disguised as artillery officers. Nolan, a radiologist, raised some suspicion. He constantly had to return below decks secretly to check the radiation from the canister, and could not answer even the most basic questions about his supposed artillery experience.

It is unclear if even Captain McVay knew exactly what his ship was carrying. *Indianapolis* travelled quickly. The cruiser stopped at Pearl Harbor briefly to refuel, before continuing on to Tinian, arriving safely on 26 July. The bomb components were unloaded and reassembled on the island. Its top-secret mission concluded, after leaving Tinian – without the crew setting foot ashore – the *Indianapolis* was ordered to Guam to rendezvous with other ships at Leyte Gulf in the Philippines. Two controversial procedural flaws on Guam would eventually contribute to the cruiser's disastrous fate.

When the cruiser sailed from Tinian, the US Navy's Pacific Command was concerned about the submarine threat in the area. Navy operations staff had two indications of dangerous submarine activity along the route to Leyte Gulf: code-breaking intercepts and the sinking of US destroyer *Underhill*. Captain McVay was not made aware of these intelligence indicators, and precautions were not taken. Large, heavy cruisers like the *Indianapolis* didn't have sonar and they normally had an escort. McVay had requested a suitably armed destroyer escort as was normal operating procedure (SOP) in an area of known submarine activity, but none materialised.

Instead, he was ordered to follow a solo zig-zag course 'at his discretion' in order to throw off enemy submarines. Before sailing, McVay checked on the tactical situation and received the following:

SECRET
PDG 1849
OFFICE OF THE PORT DIRECTOR GUAM
28 JULY 1945
/s/ J.H. Jamey
From: Port Director, Guam.

To: Commanding Officer, USS INDIANAPOLIS (CA-35).
Subj: Routing Instructions.
Encl: (A) Intelligence Report, (B) Approach Instructions, (C) Flight Brief Bulletin.
ENEMY SUBMARINE CONTACTS:

22 July: Sub sighted surfaced at 10:34N-132:47E at 0015K. Hunter-Killer ordered.

25 July: Unknown ship reports sighting as possible periscope at 13:56N-136:56E at 250800 K.

25 July: Sound contact reported at 10:30N-136:25E. Indications at that time pointed to doubtful submarine.

Commodore James Carter, head of the Pacific Fleet's advance headquarters, told McVay, 'Things are very quiet. The Japanese are on their last legs, and there's nothing to worry about.'

The last known sighting of the cruiser was by a tank landing ship in transit, whose captain later confirmed that the cruiser was making 15 knots and zig-zagging. Sometime between 19:30 and 20:00, Captain McVay determined that the visibility post-sunset was poor enough for *Indianapolis* to cease zig-zagging and increase its speed to 17 knots.

He had not reckoned with Lieutenant-Commander Mochitsura Hashimoto, who captained the Japanese submarine *I-58*; even with his nation on the brink of defeat, he was after one final trophy. As it made its way east into the Pacific, unescorted, the *Indianapolis* was spotted silhouetted against the moon. It was 30 July, just four days after the bomb had made it to Tinian. Two torpedoes hit the ship, blowing the bow open and causing secondary explosions. The ship was almost ripped in half and sank in twelve minutes. Crew members drowned, some were struck by the propellers or died in the explosions. Of the 1,196 crew, around 900 made it into the water. The men grouped together, clung to debris and sought out life rafts, but most died in the water while waiting to be rescued. Causes of death included dehydration, starvation, salt poisoning and drowning. Hallucinating men drank salt water and died. Most terrifying were the shark attacks, which came often and without warning, and are now regarded as the worst such episode in history.

Loel Dene Cox, Seaman Second Class remembered:

> You could barely keep your face out of the water. I had blisters on my shoulders, blisters on top of blisters. It was so hot we would pray for dark, and when it got dark we would pray for daylight, because it would get so cold, our teeth would chatter.
>
> In that clear water you could see the sharks circling. Then every now and then, like lightning, one would come straight up and take a sailor and take him straight down. One came up and took the sailor next to me. It was just somebody screaming, yelling or getting bit.

The problem was, no one knew that the *Indianapolis* had gone down and nobody could have known where it had happened.

As far as the commander of the Pacific Fleet was concerned, he had sent a cruiser on a top-secret mission to Tinian as ordered by the Joint Chiefs of Staff. The local admiral had then effectively hijacked a spare ship and given it orders until ComPac demanded its return. No one was really in charge of the USS *Indianapolis* when she was sunk. Her very secrecy made her an orphan of the sea.

It was not until four days later that the pilot of a routine anti-submarine patrol spotted men in the water. A seaplane commanded by Lieutenant Marks was sent to drop survival supplies, but when his crew noticed that the survivors were being attacked by sharks, he executed a dangerous open-ocean landing (in breach of orders) and taxied to pick up survivors while radioing for urgent help for his now grossly overloaded seaplane.

A destroyer, the USS *Cecil J Doyle* was first on the scene, followed by four others. Marks's seaplane had been damaged in the landing and he concentrated on getting men out of the water to safety. There were only 317 survivors, a loss unparalleled in US history with 879 crew members having perished, including 4 who died despite being rescued from the water. Questions remain about why the rescue was delayed for so long. The official version was that no distress signal had been sent, but survivors testified otherwise. Additional testimonies from other ships receiving the signal corroborate the survivors' narrative. Also, naval intelligence decrypted a signal from the *I-58* indicating that it had sunk a 'big warship' along the route that the was taking. Some reports attributed the lack of action to fears that the messages were a ploy by the Japanese, attempting to ambush rescue ships. Some historians have argued that communication about the unfortunate *Indianapolis* was lacking because of the clandestine nature of its prior mission.

But the inescapable conclusion remains that it was the secrecy around the atomic bomb and its top-secret mission that started the fatal chain of events that finally sank the *Indianapolis*.

CHAPTER 4

SOME EARLY MILITARY NEAR-MISSES

'In an operational or training setting, "negligent discharge" may be defined as the negligent discharge of a weapon caused by an act, which could and should have been foreseen and prevented. This act would be accompanied by a degree of negligence or recklessness.'
– US CODE OF MILITARY JUSTICE

Accidents happen. With nuclear weapons, that has a potentially disastrous significance. It was only a matter of time before things began to go wrong with the new weapons or their means of transport, because in human affairs things go wrong all the time.

By 1950, the United States had about 230 fully functional atomic bombs, and the Soviet Union might have had two. The spirit of the time can best be summed up by misquoting Hilaire Belloc's quip: 'Whatever happens, we have got the atom bomb and they have not.' American bases and airfields encircled the Soviet Union, and under the nuclear war plan 'Operation Dropshot', US Strategic Air Command (SAC) was planning to drop up to 300 atomic weapons on 200 Soviet and Chinese cities in the event of war. We now know that since 1950 there have been at least thirty serious nuclear incidents recorded by the United States Air Force (USAF) alone. Soviet figures remain shrouded in secrecy. In USAF terminology, an accident involving nuclear weapons is known as a 'broken arrow'. This is defined as an unexpected event involving nuclear weapons that result in 'the accidental launching, firing, detonating, theft or loss of the weapon'.

The first recorded broken arrow was the dropping of an atomic bomb in early 1950. The USAF's huge Convair B-36, with its six piston and four jet engines, was the first dedicated nuclear bomber. Entering service in 1948, the B-36 became the primary nuclear weapons delivery vehicle of the new USAF Strategic Air Command until it was replaced by new jets in the mid-1950s. With the jibe 'Two turning, two burning, two smoking, two choking, and two more unaccounted for' – the sarcastic unofficial SAC version of the multi-engined B-36's 'ready to go' capability – the B-36 was regarded as a very safe aeroplane – until the inevitable accident.

On 14 February 1950, a USAF Convair B-36, en route from Alaska to Carswell Air Force Base, Texas, was conducting a test flight over San Francisco. The aircraft carried a Mark 4 atomic bomb, containing natural uranium and 2,270 kilograms of conventional explosives, but mercifully without the plutonium needed to create a nuclear explosion. Over the US-Canadian coast, three of its engines began shooting flames, and another three engines became incapable of delivering full power (the aircraft had ten engines – six piston and four jet). The official investigation later blamed ice in the carburettor air intakes for the failures.

Struggling to keep the plane aloft, the flight engineer gave the engines more power, but three of the bomber's six piston engines caught fire and had to be shut down. To lighten the aircraft and to protect US atomic secrets, the pilot ordered the crew to release the heavy bomb, setting its fuse to detonate at 1,220 metres (4,000 feet).

In secret testimony to the US Air Force board of inquiry into the loss of the aircraft, the pilot, Captain Barry, described the final minutes of the flight: 'We were losing altitude... and I asked the radar operator to give me a heading to take me out over water. We got out over the water just about nine thousand feet [2,749 m], and the co-pilot hit the salvo switch. At first nothing happened, so he hit it again and this time it opened.'

The bomb dropped into the water in the passage between the islands and the mainland on the Pacific Northwest coast. Captain Barry's testimony takes up the story: 'The radar operator gave me a heading to take me back over land... but we were descending quickly... by the time we got over land we were down to five thousand feet [1,524 metres].'

Barry ordered the crew to bail out into the darkness. The bomber, without crew or weapon, flew on autopilot and finally crashed into the side of Mount Kologet in British Columbia, 320 kilometres to the north, although search operations by both the Canadian and US militaries found no sign of the doomed bomber at the time. Five of the crew of seventeen were never seen again and the last survivor, Lieutenant-Colonel Pooler, was found in a tree, hanging by his parachute upside down, with a broken ankle.

According to the USAF, the bomb did not contain the plutonium core necessary for a nuclear detonation and, typical of nuclear mishaps, the incident was hushed up. Four years later, the plane's wreckage was found in the Canadian wilderness, but no bomb was ever found at sea. As with the circumstances surrounding many of the twenty-three other atomic weapons that were jettisoned or inadvertently released from aircraft or ships during the first twenty-four years of the atomic age, an aura of mystery has surrounded the crash.

That early B-36 accident was to be the first of many. On 11 April 1950, three minutes after departure from Kirtland Air Force Base in Albuquerque, a USAF B-29 bomber carrying a nuclear weapon and a crew of thirteen crashed into a mountain on Manzano Base. The bomb's casing was demolished and its high explosives ignited upon contact with the plane's burning fuel. However, according to the Department of Defense, all nuclear components were recovered. A nuclear detonation was not possible because, while on board, for safety reasons, the weapon's core was not in the weapon. All thirteen crew members died.

A very similar accident took place only nine months after the B-36 crash in British Columbia. On 10 November 1950, a B-50 bomber was transporting one of eleven 'special weapons' from Goose Bay, Newfoundland, to Arizona when it lost power in two of its four engines and was forced to dump its payload over the St Lawrence River near Quebec. More than two tonnes of high explosives detonated in the air, startling thousands of *Québécois* who thought that war had begun. The Pentagon issued a cover story stating that one of their bombers with engine trouble had jettisoned a load of conventional practice bombs. The accident scattered 45 kilograms of 'ordinary' uranium (U-238) across the countryside.

The 1950s was a bad decade for the USAF and its atomic safety record. The most egregious example is perhaps the day in 1957 when the USAF bombed New Mexico. The air force kept the nuclear accident secret for twenty-nine years. The story is one of near tragedy and pure farce.

Released government documents revealed that a 19,050-kilogram Mark 17 hydrogen bomb fell from an Air Force B-36 bomber flying from Biggs Army Air Field in Texas on 22 May 1957. The giant bomber was commanded by veteran pilot Lieutenant-Colonel Richard Meyer. Standard operating procedure was the removal for take-off and landing of the safety locking pins that ensured the bomb could not fall out accidentally during flight. This awkward job had to be done by hand and required a crew member, usually the navigator, to climb into the bomb bay to insert and later remove the large metal safety pin. On that day, First Lieutenant Bob Carp was assigned the tricky task.

With the plane descending to 518 metres (1,700 feet) and on its final approach before landing at Kirtland Air Force Base, Carp began the difficult job, hanging on with one hand in the bomb bay while leaning over the 8-metre long, steel-encased weapon to retrieve the pin.

The plane was about six kilometres south of the airfield, and landing conditions were normal as Carp reached out for the pin. What happened next is in dispute. The huge bomber bounced in a pocket of turbulent air. Carp grabbed for the nearest hand-hold, a lever that immediately gave way under his weight, loosening the huge bomb from its shackles to smash through the closed bomb doors, ripping them away. Released from its 20-tonne payload, the bomber shot up by some 400 metres. In the cockpit, the startled pilot reacted with, 'Oh shit!'

As the only eyewitness to the events in the bomb bay, Carp was emphatic that a 'defectively designed' manual release mechanism had been accidentally pulled into release mode as he grabbed for the safety pin.

'Bombs away!' screamed a nearby crewman. According to another witness, Carp, his face 'whiter than any sheet you ever saw', slowly pulled himself out of the remaining bomb bay, yelling, 'I didn't touch anything.' The B-36 promptly sent a distress call to the Kirtland tower reporting, 'We've dropped a hydrogen bomb!'

The bomb itself plummeted downwards but the 500-metre drop was far too short for its parachutes to deploy. Long before the plane could pull away, the weapon smashed into the ground. When it impacted, six kilometres south of Kirtland Air Force Base's control tower, the non-nuclear explosives used to trigger nuclear devices detonated. Fortunately, no one was injured as the bomb hit an uninhabited area owned by the University of New Mexico.

The subsequent investigation conducted recovery and clean-up operations at the site shortly after the accident and declared no nuclear hazard. Only the bomb's conventional explosives had been triggered by the fall. 'Minor radioactive contamination' was detected in the crater and no radioactivity was reported beyond the lip of the crater. On 5 August, there was another non-nuclear detonation of an atomic bomb at Fairfield-Suisun Army Air Base in California. A USAF B-29 bomber flying a Mark 4 nuclear bomb to Guam experienced malfunctions during take-off and crashed while attempting an emergency landing. In the resulting fire, the bomb's high-explosive material exploded, killing nineteen people from the crew and rescue personnel.

Two more USAF nuclear near misses tell us that things did not improve with time. In 1956 at RAF Lakenheath in England, a B-47 Stratojet carrying a thermonuclear bomb came within a whisker of disaster. The aircraft was practising touch-and-go landings on return from a mission. On the fourth pass, something went wrong, and it ploughed into a nuclear-bomb storage bunker, or 'igloo', bursting into flames. Three Mark 6 nuclear bombs were in storage, although without their nuclear detonators which, for safety reasons, were stored elsewhere. Yet each Mark 6 contained 2,270 kilograms of high explosives, and depleted uranium. Even if the weapons had detonated because of the flames, there would not have been a nuclear reaction as uranium-238 is not fissionable through explosion or fire.

Part of an official USAF signal about the incident, marked 'Top Secret', says it all:

> A/C [aircraft] then exploded showering burning fuel overall. Crew perished . . . Preliminary exam by bomb disposal officer says a miracle that one Mark Six with

exposed detonators sheared didn't go [i.e. detonate]. Firefighters extinguished fire around Mark Sixes fast.

Human error figures large in this story of nuclear blunders. In 1958, a very similar story pointed the finger unerringly at a bungling bombardier as the culprit when another nuclear bomb was unloaded by the USAF on unsuspecting civilians back home. A USAF B-47E Stratojet bomber was flying from Savannah, Georgia, to England on a mock bombing exercise. The flight navigator/bombardier was told to go into the bomb bay and check the locking harness on the massive, 3,450-kilogram Mark 6 nuclear bomb. Scrabbling around, he accidentally pushed the emergency bomb release lever – with disastrous results.

The bomb shackles opened and the bomb crashed through the bomb-bay doors below, smashing the thin metal open, just as Bob Carp had experienced the previous year in a B-36. This time, however, the bomb went into a 4,500-metre free fall. The detonator for its conventional high explosives went off after it hit the ground 10.5 kilometres east of Florence, South Carolina, creating a 20-metre wide crater, 9 metres deep. A nearby house was destroyed and several people were injured. History does not relate what happened to the bungling bombardier, but the accident joined the long list of the USAF's broken arrow atomic discharges during the 1950s.

The UK was not blameless, although British official records show only two major incidents involving the RAF's much smaller nuclear-armed 'V Force'. One of them occurred when a nuclear bomb simply fell out of a V-bomber's bomb bay and dented itself on the tarmac, to the shock of the watching ground crew. One later said, 'We just stood there, holding our breath.'

The second incident, on the night of 8 August 1967, was more serious. A major thunderstorm rolled across RAF Waddington in Lincolnshire while there were loaded V bombers parked in the base's ground alert area – Vulcan bombers at 'Quick Reaction Alert', able to take off within two minutes in the event of war to spearhead a nuclear attack on the Soviet Union. The ground staff looked on aghast as one of the nuclear-armed planes was struck by lightning. 'It was a bit like a firework which you have lit and it has not gone bang,' one of the airmen later commented. However, although it had been scorched by the flash and some electronics had been burnt, the Vulcan did not catch fire or explode and the nuclear bomb it carried had not been damaged.

That British incident highlights that even the simplest and most mundane event or task has inherent risks whenever nuclear weapons are involved. On 13 November 1963, a careless forklift driver at Medina Base Annex in Texas accidentally scraped his truck along a load of weapon components containing high explosive, but no plutonium. They caught fire and blew up, 55,800 kilograms of high explosives vaporising the storage igloo and hurling weapon components across the countryside. Despite early fears, the Medina complex, Lackland Air Force Base and San Antonio, Texas, did not become radioactive wastelands.

One of the difficulties we have in assessing the number and dangers of these early nuclear accidents is the lack of information about the USSR and its early nuclear efforts. Logic dictates that they too must have had their fair share of mishaps. We do know that Soviet dictator Joseph Stalin authorised a major Soviet atomic bomb project to research and develop nuclear weapons during what Russians call the 'Great Patriotic War' (describing their country's involvement in World War II, from the Nazi invasion of Soviet Russia in 1941 to Germany's capitulation in 1945).

The Soviet scientific community was well aware of the potential of an atomic bomb. Independent research on radioactivity was being conducted as early as 1910 by several

Russian scientists. By 1939, an alert Russian physicist, Georgy Flyorov, noticed that learned articles and papers in scientific publications about nuclear fission by the usual German, American and British scientists, were suddenly drying up. In an echo of Einstein's letter to President Roosevelt, Flyorov realised the significance of this silence and wrote directly to Stalin, urging him to develop a Soviet nuclear bomb and declaring: 'The results will be so overriding [that] it won't be necessary to determine who is to blame for the fact that this work has been neglected in our country.' Flyorov and fellow physicist Konstantin Petrzhak emphasised: 'It is essential to manufacture a uranium bomb without a delay.'

The Soviet leader directed a full-scale programme. Although the German invasion of western Russia in 1941 and 1942 slowed the programme down, the USSR benefited from the wealth of nuclear intelligence flooding in from their spy rings, many relying on traitors and idealists among those working on the US Manhattan Project, such as Klaus Fuchs and Britain's Tube Alloys.

After Stalin learned of the 1945 atomic bombings of Hiroshima and Nagasaki, the programme was accelerated. Ironically, when Truman told Stalin about the new US atom bombs at the Potsdam Conference in July/August 1945, the Soviet leader just grunted and nodded. He was almost certainly better briefed on US nuclear matters than the American president. That same year, Soviet intelligence even obtained blueprints of the first US atomic device. The result was that on 29 August 1949, the Soviet Union secretly conducted its first successful weapon test of 'First Lightning', based on the American 'Fat Man' design, at the Semipalatinsk Test Site in Kazakhstan. The American monopoly of nuclear weapons was now lost for ever.

That the Soviets subsequently had a string of atomic accidents is without question. Details are scant, but one serious nuclear disaster demonstrates the USSR's cavalier approach to safety.

A secret nuclear test in 1954 went unreported until the Soviet newspaper *Pravda* revealed the details in 1991. Two years later, the episode was recounted in the Finnish documentary *Human Nuclear Guinea Pigs*.

In September 1954, as part of a military exercise, the Soviet military detonated an atomic bomb near the village of Totskoye in the Ural Mountains, within three kilometres of 45,000 troops and several thousand civilians. The aim of the exercise was to test whether troops could fight a battle in an area immediately after a nuclear bomb had hit. The military's cameras recorded the explosion as it happened, 1,000 feet (350 metres) in the air, with the mushroom cloud forming almost directly above troops hunkered down in makeshift shelters.

Up close, the film showed images of flaming houses, scorched animals and mangled military equipment. Soon after the shock waves subsided, troops with little or no protective gear were filmed storming through an inferno of dust, heat and radiation.

About one million people lived within 160 kilometres of the detonation at the time, but the numbers of deaths, injuries and illnesses resulting from the incident are unknown. A group of veterans said they had experienced radiation illness for years following the incident, at least until the 1990s. Villagers nearby climbed on their roofs to watch the explosion – some went blind, and others later developed cancer. US intelligence reports compared the explosive power of the bomb to the one that the US dropped on Hiroshima – about 16 kilotons. But the film showed that the Soviets were prepared to expose their soldiers to radiation levels ten times higher than the maximum then permitted for a whole year for American troops.

Totskoye demonstrated the Soviets' willingness to push nuclear safety to the limit in the early days. The Soviet Union, inefficient and careless as ever, had a very bad record for keeping track of its nuclear devices. The problem is that so many of these accidents behind the Iron Curtain remain undocumented in the

West, although a little bit of reverse logic and precedent can guide us on early Soviet nuclear accidents.

It is known that the Kremlin has admitted fifteen nuclear accidents between 2000 and 2010, a series that could act as a template for the kind of serious incidents which may have happened fifty years earlier, in the 1950s. This more recent list of nuclear and biological accidents is instructive:

1. June 2000 – Servicemen at a naval base in the Russian Far East were poisoned after a nuclear missile leaked fuel.

2. 12 August 2000 – An explosion sinks the *Kursk* nuclear-powered submarine, killing all 118 crew on board.

3. 30 August 2003 – Nine crew members die when the decommissioned *K-159* nuclear submarine sinks in the Barents Sea, while being towed.

4. 5 May 2004 – A researcher at a Siberian biological laboratory pricks herself with a needle that is found to be contaminated with the Ebola virus, and subsequently dies.

5. 5 August 2005 – With British help, seven Russian sailors are rescued after becoming trapped in the *AS-28* deep-submergence rescue vehicle, which in turn had become trapped in underwater nets.

6. 7 September 2006 – There are two fatalities after a fire breaks out on the *Daniil Moskovskiy* submarine as it is towed across the Barents Sea.

7. 15 July 2008 – In Chechnya, seven soldiers die and a further six are injured when a tank shell explodes.

8. 17 October 2008 – When a MiG-29 crashes in the southern region of Siberia, the fleet has to be grounded for a time.

9. 8 November 2008 – Twenty people are killed and at least twenty-two more are injured when the Freon fire-extinguishing system on board the *K-152 Nerpa* submarine is activated.

10. 5 December 2008 – In southern Siberia, a MiG-29 loses some of its tail section due to corrosion and crashes. As the whole of the fleet is inspected, some ninety planes are found to be unsafe.

11. 7 January 2009 – After a fire breaks out on the *Admiral Kuznetsov* aircraft carrier just off Turkey, one crew member perishes from carbon monoxide poisoning.

12. March 2009 – The decommissioned nuclear submarine *Orenburg* catches fire at a naval yard in Severodvinsk.

13. October 2009 – A further fire occurs at Severodvinsk – this time while the submarine *Kazan* is being decommissioned.

14. 14 November 2009 – After an explosion at the navy nuclear depot Arsenal 31, near Ulyanovsk, two fire-fighters die during the decommissioning of munitions.

15. 23 November 2009 – A second explosion occurs at Arsenal 31, with eight soldiers losing their lives.

From this catalogue of admitted military accidents, we can infer that the USSR must have suffered as many accidents as the USAF of the day. The Soviet submarine force's nuclear safety record in particular is appalling. The first Soviet nuclear boat, *K-19*, suffered so many defects and accidents that its nervous crew nicknamed it 'Hiroshima'. The submarine had been designed and built by the Soviets in a rush, in response to the United States' nuclear advances. It was badly designed and hastily constructed. Before it was even launched, ten civilian workers and a sailor had died owing to accidents and fires.

After it was commissioned, *K-19* had multiple breakdowns and accidents, several of which threatened to sink the submarine. During its initial voyage on 4 July 1961, the coolant for the nuclear reactor failed, killing twenty-two members of the crew. The submarine later experienced numerous other defects and accidents, including two fires and a collision.

While the CIA and the Office of Naval Intelligence had a pretty good idea of the floating disaster that *K-19* represented and the Soviets' nuclear blunders, both they and the Kremlin kept a tight lid on the truth – and still do. One cannot help but wonder how many more serious incidents have been suppressed and covered up under the catch-all Russian blanket of national security.

Overall, the suspicion must be that the nuclear accidents about which we know may represent the tip of an iceberg. Other sources throw doubt on the bland official versions of events. 'Playing With Fire', a 2017 report by an independent organisation known as the Nuclear Information Service, presents the accident record of the UK's nuclear weapons programme over its sixty-five-year history. Among other revelations, it discovered that:

- Only on one occasion, in 2003, has the Ministry of Defence (MOD) released an official list of accidents involving the UK's nuclear weaponry.
- The twenty-seven incidents do not represent a full tally of all such mishaps.
- It enumerates 110 accidents, near-misses and dangerous events that have occurred during the history of the UK's nuclear weapons programme. Among them are fourteen significant accidents that occurred during nuclear-weapon production, and twenty-two related to transporting nuclear weaponry on roads.

- The storage and handling of nuclear weaponry account for eight incidents.
- There have been forty-five accidents involving nuclear-capable aircraft, ships and submarines.
- A further twenty-one events involving security failings.

The Campaign for Nuclear Disarmament (CND) separately identified the first 'serious incident' at RAF Greenham Common as early as August 1957. At the time, a UK Ministry of Defence report maintained that the three-week closure of the base was to contain an outbreak of 'Asian flu'. It must have been quite a nasty outbreak: eighty-three US personnel were hospitalised on the base; barracks were built to house them. The authorities at Greenham Common told the local *Newbury Weekly News* that an 'emergency' had closed the base, resulting from an 'accidental spill of 2,000 gallons [7,570 litres] of fuel in the neighbourhood of six aircraft'.

However, in 1996, CND uncovered new documents at the UK Public Record Office describing an unreported incident at Greenham Common in 1957, involving a fire on a US B-47 nuclear bomber, which burned while its weapons were still inside.

British scientists at Aldermaston's Atomic Weapons Research Establishment were sceptical. Dr F D Morgan, a senior scientist at Aldermaston, wrote to Professor Sir William Penney, who was Britain's foremost weapons expert at the time of the incident, revealing his worries about the fire. He and his team had been alerted to contamination levels of isotope uranium-235 that were around a hundred times higher than usual in the vicinity of the base. In his note to Penney, he stated that he believed nuclear material had been released at Greenham, and that the MOD had not been told by the US Air Force. A ministry spokeswoman in 1996 stated flatly: 'There has never been an accident involving damage to a nuclear weapon in the UK. At the most we are talking about scratches. Somebody might have dropped one on the ground, and it is classified as an accident.'

When the report became public in 1996, both CND and the opposition parties of the time pushed for the complete story of events from the MOD. 'Each line of defence,' they said, 'looks more implausible than the last. If the MOD is right in claiming there have only been minor incidents then they have nothing to lose by publishing the details, but they refuse to do that. The MOD should have learned ... that a cover up becomes a bigger story in the end.'

The worrying conclusion from this litany of nuclear military mishaps is that there have simply been far too many over the years. Sooner or later, one will go very seriously wrong, with incalculable consequences. To make matters worse, governments invariably lie when confronted with potentially embarrassing allegations of official blundering – particularly the military, hiding behind their catch-all cloak of 'national security'. It's in their nature and it certainly does not inspire confidence in our governments' control of matters nuclear.

Up to now we've been lucky.

CHAPTER 5

BALLS-UP AT BIKINI

'The sight of the first woman in the minimal two-piece was as explosive as the detonation of the atomic bomb by the US at Bikini Island.'
– TOM WAITS, MUSICIAN

Scientists, like doctors, like to pretend they know the answers. This is a conceit – after all, it was the medical profession that, for thousands of years, told their patients that bleeding was good for wounds, cupping could cure many internal ailments and frontal brain lobotomies were splendid for mental health. And we must never forget that from 1945 to 1947, Manhattan Project doctors secretly injected eighteen unwilling or unknowing victims with plutonium, as directed by the United States Atomic Energy Commission. One even compared what he was being told to do as being no different from the practices of Dr Mengele. Truly, Voltaire was right when he said of doctors, 'The art of medicine is keeping the patient amused while nature tries to effect a cure.' Or, in the case of the American plutonium injections, 'Let's see if this does some harm.'

Scientists are no more omniscient than medics, however

much they would protest otherwise. History is replete with curious scientists who got it wrong and chased scientific progress down some blind alley of knowledge. Arguably their biggest blunder, in the true sense of the word, was a massive nuclear miscalculation – Bikini Atoll and the first hydrogen bomb.

The idea of a thermonuclear fusion bomb detonated by a smaller fission bomb was originally mooted by Enrico Fermi to his colleague Edward Teller in 1941. Teller became obsessed by a *fusion* bomb at the expense of atomic *fission* (which was what he was being paid to do on the Manhattan Project) and actually spent more time trying to work out his big idea than sorting out the details of the original atomic bomb. Scientific Director J Robert Oppenheimer left the distracted Teller to himself.

Hydrogen bombs are extremely complicated devices. They are really two bombs in one. The first is an old-fashioned atomic bomb, 'the primary' – a standard atom fission bomb with a normal chain reaction but set to explode deep inside another more destructive, yet difficult-to-detonate nuclear device, 'the secondary'. The explosion of the first bomb starts a major reaction in the surrounding shell of plutonium and volatile uranium-235 by blasting it with four types of energy: expanding hot gases; superheated plasma; electromagnetic radiation; and a bombardment of free neutrons from the primary's nuclear detonation looking to combine with other atoms. It is this booster that sets off the main hydrogen explosion.

The fission energy released in the primary 'sparkplug' begins a chain reaction, and induces lithium deuteride, an isotope of hydrogen, to go critical. This pumps out escaping neutrons to bond and react with everything in sight, as escaping neutrons do, especially with tritium, an isotope of hydrogen that already has two neutrons, and therefore allows for far more efficient fusion reactions to occur during the detonation of a nuclear weapon. This in turn sets off the main hydrogen explosion, as the whole chain reaction then goes critical and an uncontrolled nuclear explosion follows. The two processes can be graphically represented as follows:

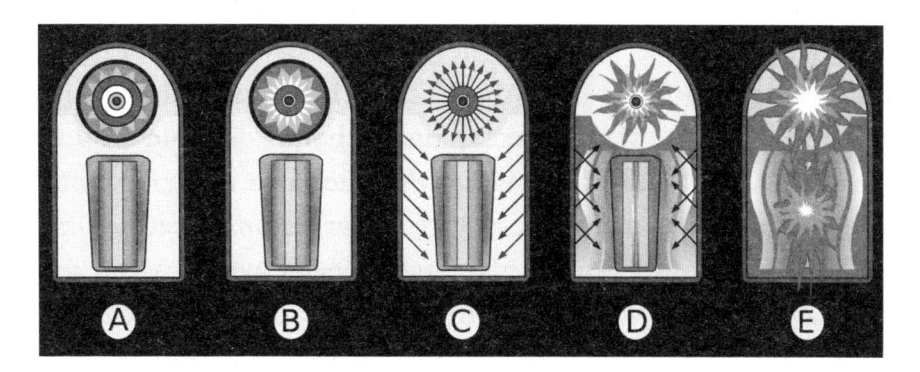

A Warhead before firing; primary at top.

B High-explosive fires in primary, compressing plutonium core into supercriticality and beginning a fission reaction.

C Fission in primary emits X-rays, which channel along the inside of the casing, irradiating the polystyrene foam channel filler.

D Polystyrene foam becomes plasma, compressing secondary, and plutonium sparkplug inside the secondary begins to fission, supplying heat.

E Compressed and heated, lithium-6 deuterium fuel begins fusion reaction, neutron flux causes tamper to fission. A fireball starts to form.

Illustration adapted from Howard Morland, *The Secret That Exploded* (New York, Random House, 1981).

The key is the link between the primary and secondary explosions, B and D, the all-important interstage (*see* illustration). This is the most highly classified part of the hydrogen bomb, also referred to as the 'H-bomb'. In the early days, it was literally a matter of speculative science and informed guesswork. Some of the materials used have never been identified or confirmed and are hidden behind the aptly named codeword 'Fogbank'.

From this basic technical description of the two-stage process needed to initiate a hydrogen bomb, it is clear that the nuclear physics and engineering required are far from simple. A great deal can go wrong.

Teller's work bore fruit: the US military eventually backed his programme. Meanwhile, straightforward atomic testing continued at the remote Bikini Atoll in the South Pacific. 'Operation Crossroads', which took place on 1 and 25 July 1946, was the first test of America's new nuclear weapons since the 1945 attacks on Japan. The trial was intended to advance scientific knowledge about atom bombs and radiation, but it had more far-reaching effects. The plan, led by the US Navy – desperate to get in on the atomic act – was to carry out three nuclear explosions but they were not entirely successful. The main aim was to discover what happened to naval warships when a nuclear weapon exploded. Much to the fury of animal rights protesters, some of the ships were loaded with live animals, such as pigs and rats, in order for scientists to study the effects of the nuclear blast and radioactive fallout on them. More than ninety vessels, (not all carrying live cargo) were placed in the target area of the bomb, which was dropped by a B-29 bomber from 9,145 metres (30,000 feet).

Unfortunately, the first bomb was off target by 800 metres (half a mile). Embarrassingly, the 'weapon exploded almost directly above the navy's data-gathering equipment, sinking one of its instrument ships, and a signal that was meant to trigger dozens of cameras was sent ten seconds too late,' one observer wrote. The real consequence of these botched tests was to establish remote Bikini Atoll as the home for more American nuclear tests. In all, twenty-three nuclear trials were conducted at Bikini, including the first test of an H-bomb in 1954, the largest nuclear device the US ever exploded. This was not an unalloyed success. That H-bomb test at Bikini turned out to be a major scientific balls-up.

Known as 'Castle Bravo', it was calculated to verify design concepts that the scientists hoped would then become the standard for all future US high-yield thermonuclear weapons. In this, 'Castle Bravo' succeeded, but it was also the scene of one of the worst early nuclear accidents ever – and one kept very quiet for decades.

The problem was that the experts completely underestimated the real power of the early hydrogen bombs. Anyone who has seen the film of this bomb blast would have been impressed by its size – and alarmed. Because on 1 March 1954 at Bikini Atoll in the Marshall Islands, the physicists and mathematicians got it badly wrong. The bomb's yield of 15 megatons turned out to be two and a half times the 6-megaton maximum predicted by its designers.

It transpired that the various nuclear isotopes, and particularly the new lithium-tritium mix, yielded far more energy and radiation than the scientists had calculated. Of the total yield, 10 megatons came from fission of the natural uranium tamper – a much higher yield than was originally expected. Combined with the far larger-than-expected yield from the secondary fusion part of the bomb, the cumulative blast was massively greater than predicted. Worse, an unanticipated wind shift caused the greater radioactive fallout to spread into unexpected areas hundreds of kilometres downwind. This unexpected blast, a thousand times stronger than that of the Hiroshima bomb, gave rise to widespread radioactive fallout that affected Australia, India and Japan, not just a few isolated Pacific islands and their unsuspecting inhabitants. The Americans' Bikini H-bomb test turned into an international disaster.

First to suffer were some unfortunate Japanese fishermen. In those days, their fishing vessels had no radios, so no one could warn them that they would be near an area where an H-bomb was to be detonated. Just before dawn on 1 March 1954, about 160 kilometres (100 miles) east of Bikini, a Japanese fishing boat called the *Daigo Fukuryū Maru* ('Fifth Lucky Dragon') bobbed on the ocean with most of the crew below deck asleep. They were abruptly woken, as recorded in the log of Yoshio Masaki, the ship's fishing master: 'Suddenly the boat has been surrounded by a bright light. Such an early dawn is impossible. Makes feel something very dangerous.'

What they had witnessed was far more than a Hiroshima-style atomic bomb. The glow came from the test detonation of that first thermonuclear H-bomb, mankind's newest and most powerful tool of war. Thanks to the scientists' miscalculations, the unlucky crew of the *Fifth Lucky Dragon* were well within the range of the new H-bomb's impact. As the crew went to work hauling in their overnight catch, it started to rain. Sudden howling winds – which the US meteorologists had predicted would blow the other way – brought monsoon-like precipitation, covering the ship and crew with a blanket of white, gritty ash.

The *Fifth Lucky Dragon* was sailing directly through the radiologically contaminated zone. The crew had no idea what the sediment was, but their boat was soon covered with it, so they got out their brooms and swept it off. They kept picking it up and looking at it, trying to figure out what it was and where it was coming from. The rain and ash fell on the *Fifth Lucky Dragon* for five hours.

By the time it passed, some of the crew were dizzy, vomiting or feverish. They had been covered in, swallowed and inhaled the highly radioactive remains of a large chunk of Bikini Atoll's coral, incinerated to dust by the immense nuclear explosion. It had then rained down across a vast area of the Pacific. By the time they got back to port two weeks later, the whole crew had to be hospitalised. The *Yomiuri Shimbun* newspaper reported 'Japanese fishermen encounter Bikini A-bomb explosion test. 23 men suffer from A-bomb disease.' Within days the international press was covering the event, too.

Japanese medical investigators found that the men of the *Fifth Lucky Dragon* were suffering from something other than A-bomb sickness, Hiroshima-style. They were afflicted with acute radiation disease, caused not by the bomb but by the radioactive rain it produced. The Japanese called this rain *shi no hai*, or 'death ash'. The media, and the world, quickly found a new name to describe this fearful new phenomenon: 'radioactive fallout'.

Ever since 1945, Japan had tended to sideline and downplay the *hibakusha*, the survivors of the atomic bombs at Hiroshima and Nagasaki, but now sympathy for the *Fifth Lucky Dragon* crew poured out, backed by outrage at the United States for once again victimising Japan with atomic weapons: 'We are not guinea pigs!' wrote one newspaper. The test resulted in an international uproar and reignited Japanese concerns about nuclear radiation, especially with regard to the possible contamination of fish.

According to a diplomat in the economics section of the US Embassy in Tokyo at the time (clearly mostly concerned with the financial implications of the disaster):

> That fishing boat went chugging on back to port with a sick crew and a hatch full of fish. When they got to port the catch was unloaded and put into the Japanese distribution system. The Japanese made the mistake of letting that catch be distributed throughout the country ... they didn't know where the fish had gone, they lost track of distribution. Even in Tokyo the enormous fish market sold very few fish for weeks.
>
> Then the crewmen began reporting to the hospital. And then they started demanding compensation, of course. Two of the crewmen died. One of them was brought up to Tokyo to be hospitalised where he was given blood transfusions which it later became clear gave

him the hepatitis that killed him. He probably didn't die of radiation sickness at all. We in the embassy were jumping up and down and the United States was jumping up and down, because the Japanese refused to allow him to be examined by American physicians. They were demanding enormous compensation from us in various forms but were not allowing us to have any part in the treatment.

But that was just the start of the USA's problems. It soon became apparent that the test had created a much more serious nuclear disaster, one that involved more than just a few unlucky Japanese fishermen who had unwittingly sailed into harm's way. The fallout from the huge explosion settled on remote islands hundreds of kilometres away, carried by the strong winds. These islands had not been evacuated before the explosion, owing to the unanticipated fallout zone and the financial cost involved, but many of the islanders quickly began to exhibit early symptoms of radiation sickness. Those who suffered worst were the island populations of Rongelap, only 190 kilometres downwind to the east.

Many of the Rongelap islanders would suffer burns, cancers, birth defects and other medical tragedies as a result of radiation poisoning. Their first hasty exit occurred only two days after the 'Castle Bravo' test caused unexpected widespread contamination of their island home. The exiles returned in 1957 with assurances that the radiation levels on their atoll were safe. But in 1985, the islanders became convinced that radiation contamination on Rongelap was still contributing to medical problems and again moved from their home, this time to exile on the island of Mejatto near Kwajalein Atoll. Many of the Marshallese were finally resettled on other Pacific islands or in the United States. They and their descendants still cannot return to Bikini, which remains contaminated by radiation.

The H-bomb test at Bikini was not the last major nuclear test to go badly wrong, and a serious nuclear reality was soon to be revealed: *bombs could be too big*. The problem was that the Soviets were frightened by the Bikini bomb yield and decided to respond in kind – but on a greater scale. In August 2020, Russia released the truth about their monster bomb in a declassified documentary video of the 'Tsar Bomba' test on YouTube.

It appeared that the Soviets had a nasty surprise in 1961 after their nuclear scientists had oversold what became known as the 'Tsar Bomba'. This was the most physically powerful nuclear device ever deployed on earth, according to later estimates.

The prime motive for trying to build 'the biggest bomb' appears to have been for the Kremlin's political and propaganda purposes and as a response to the United States' nuclear deterrence capabilities. Premier Khrushchev's aim was to 'guarantee retaliation with an unacceptable level of damage to the enemy' in the event of a nuclear strike on the USSR – in short, to warn the Americans off. This 'Malenkov-Khrushchev nuclear doctrine' required a powerful challenge to the United States in the nuclear arms race, but 'in a distinctly asymmetrical style'. The plan was to be 'able to destroy vast areas and destroy whole cities and entire urbanised regions in one strike' (i.e. with a single missile payload or a single aircraft).

The plan for a superbomb was finally approved by the Soviet Council of Ministers on 23 June 1960 and centred around a new N-1 orbital combat rocket. With a vast new nuclear warhead weighing 70 tonnes, its estimated nuclear yield was planned to be 50 megatons (and that was half what was first intended, the scientists having balked at the potential radioactive fallout of 100 megatons).

This was a quantum jump in nuclear yield; for comparison the US largest nuke up to then had a predicted maximum yield of 25 megatons. The largest nuclear device ever physically tested by the US, ('Castle Bravo') yielded 15 megatons of TNT, and the largest weapons deployed by the Soviet Union up to then had been at around 25 megatons.

Events did not turn out as planned. On 30 October 1961, a specially modified Tu-95 strategic bomber, painted protective anti-flash white, dropped the 'Tsar Bomba', by giant parachute, over Novaya Zemlya, a remote and barely inhabited archipelago in the Barents Sea. The bomb was designed to drift slowly down to a predetermined height – 3,940 metres (13,000 feet) – and then detonate. By then, the Tu-95 and a second aircraft, carrying cameraman along with other observers, would be some fifty kilometres away; far enough away for them and their occupants to survive.

The bomb detonated at an altitude of 4,200 metres (13,780 feet). The unprecedented explosion was expected to measure about 51.5 megatons: in fact, it yielded an estimated 58.6 megatons. The fireball alone was nearly 5 kilometres (3 miles) across and was visible from a distance of 1,000 kilometres (over 620 miles). The mushroom of the explosion rose up to 67 kilometres (42 miles) and had a total diameter of 95 kilometres (59 miles). For an hour after the explosion all radio waves were disrupted by the massive EMP and total disruptive ionization of the atmosphere. The blast wave circled the planet three times. Eight hundred kilometres away, an Arctic test station had its windows blown out.

Closer to home, the effects were catastrophic. In the (uninhabited) village of Severny, on Novaya Zemlya some 55 kilometres (34 miles) from ground zero, all buildings were obliterated, while in Soviet districts hundreds of kilometres from the blast zone, damage of all kinds was reported – doors and windows crumbled or shattered, roofs collapsed, houses fell down; inhabitants were terrified. The heat from the explosion transformed the calm snowy landscape of ground zero to a charred wasteland.

The power of the explosion exceeded *tenfold* the combined

power of all conventional explosives used by all countries during World War II. The explosive energy released exceeded 3,800 times the 'Little Boy' bomb dropped on Hiroshima – 1,500 times the combined power of the bombs – 'Little Boy' and 'Fat Man' – dropped on Hiroshima and Nagasaki.

This could not, of course, be kept secret. The US had a spyplane over the Arctic code-named 'Speedlight', and its data was passed to the Foreign Weapons Evaluation Panel to analyse. US scientists were shocked. International condemnation soon followed, not only from the US and Britain, but from the USSR's Scandinavian neighbours, such as Sweden. In one Russian analyst's words, 'The only silver lining in this mushroom cloud was that, because the fireball had not made contact with the Earth, there was a surprisingly low amount of radiation.'

It could have been very different. But for a change in its design to rein in some of the power it could unleash, 'Tsar Bomba' was supposed to have been *twice* as powerful. The nuclear blast shook more than the planet. Soviet scientists – and later Western analysts – realised that the 'Tsar Bomba' 1961 bomb was just too big to use in war. It was impractical; worse, it was bloody dangerous. This dawning realisation had far-reaching effects of a very different kind.

Even Sakharov, the monster bomb's designer, voiced serious misgivings. He was worried by the amount of radioactive carbon 14 – an isotope with a particularly long half-life – that was being emitted into the atmosphere. 'This has been partly mitigated by all the fossil fuel carbon in the atmosphere which has diluted it,' he said. 'Otherwise...'

Sakharov also pointed out that a bomb bigger than the one tested would not be repelled or contained by its own blastwave – as 'Tsar Bomba' had been – and would cause global fallout, spreading toxic dirt across the planet. (Sakharov, 1921–1989, today best known as a dissident, was later to be awarded the Nobel Peace Prize.) The Soviets quietly shelved the idea of a massive bomb to end all bombs. Concern over the test hastened the end of atmospheric testing. No bomb matching its power has ever been tested again. In 1963 the United States, the USSR and the United Kingdom signed the Limited Nuclear Test Ban Treaty, which prohibits airborne nuclear weapons tests.

The Soviet experiment spooked the Americans however, and in 1962, at the height of the Cold War, the USA – worried that a Soviet nuclear bomb detonated in space could damage or destroy its intercontinental missiles – launched a Thor missile armed with an H-bomb southwest of Hawaii. Code-named 'Starfish Prime', the Thor was part of a dangerous series of very high-altitude nuclear bomb tests to discover what happens when nuclear weapons are detonated in space. When 'Starfish Prime's' 1.4-megaton nuclear warhead detonated at the pre-programmed height of 380 kilometres (236 miles) above the atmosphere, the consequences were devastating. Its immediate effects were seen for thousands of kilometres in the form of a huge aurora round the Pacific.

The other effect was a massive electro-magnetic pulse (EMP), which shocked scientists and engineers; 'Starfish Prime's' pulse was far larger than expected. In far-away Hawaii, the EMP overloaded and blew out hundreds of streetlights and caused widespread telephone outages. Other effects included electrical surges on planes, overloaded electric circuits and transistors burning out and radio blackouts. Another effect unforeseen by the scientists was the massive burst of electrons from the detonation, which damaged at least six satellites (including a Soviet one), all of which eventually failed, at a cost of billions of dollars.

CHAPTER 6

THE FRENCH FOUL-UP

'A great country worthy of the name does not have any friends.'
– **CHARLES DE GAULLE**

In all fairness, we cannot blame America's scientists alone for all the nuclear problems of the South Pacific. The French managed to cause just as much grief, if not more, to the sorely tried inhabitants of paradise.

Just as the Americans were beginning to realise that atmospheric testing of nuclear weapons was a dangerous and thoroughly bad idea, the French began a series of atomic tests using their Pacific colonies as the base. France was the last country in the 1950s to develop a nuclear testing programme, driven by the arrogance and hubris of its imperious leader, Charles de Gaulle. When other countries were negotiating to end nuclear proliferation and atmospheric testing, France's president was determined to make his country an independent nuclear power, whatever the cost or the consequences.

But that was classic de Gaulle, He was a complex man, driven by three main factors: an extraordinary and arrogant self-belief that he was a 'man of destiny', something he had openly talked about since his early officer cadet days at the Saint-Cyr military academy; second, an unshakeable belief in France as an imperial global power; and finally, a burning sense of shame and bitterness at the country's defeat in 1940 and his subsequent subservient role in World War II. To a man convinced that it was his personal mission to make France great again, possession of nuclear weapons was a vital symbol of global power.

De Gaulle's allies detested his arrogance and conceit. His habit of maintaining a proud and haughty demeanour at all times didn't help. Having fled to Britain after the collapse of the French army in 1940, de Gaulle – then still only a minor two-star general – was sentenced to death by the Vichy regime, France's puppet government under Nazi rule. De Gaulle cast himself as the embodiment of the French nation in exile, a modern-day male Joan of Arc who would lead the fight against the Germans and their Vichy hirelings, restoring France to its rightful place and greatness.

Churchill, when asked for his opinion of Charles de Gaulle, pondered, 'Do I regard de Gaulle as a great man? Let us see; he is selfish, he is arrogant, he believes he is the centre of the world . . . You are quite right. He is a great man.' The dislike was mutual. One of de Gaulle's wartime advisers noted, 'The General must constantly be reminded that our main enemy is Germany. Nazi Germany. If he would follow his own inclination, it would be England.'

President Dwight D 'Ike' Eisenhower not only disliked the leader of the Free French, but distrusted him. In a 1960 letter, Ike actually wrote to de Gaulle, by then President of France, 'I must confess, my dear General, that I cannot quite understand the basic philosophy of France today . . .' The answer was simple: de Gaulle, in his own words, saw himself as France incarnate.

In the 1960s, that meant only one thing to de Gaulle – the politics of grandeur and a determination that France as a major power must not rely on other countries for its national security and defence. Even though the USA was providing a solid deterrent against attack and the French part of NATO, which gave it nuclear protection, de Gaulle was determined to go it alone. The French did not wish to be left out of the nuclear club or to be dependent on NATO, both of which would diminish France's independent ability to deter attack through military power, let alone de Gaulle's ego. In 1966, he abruptly withdraw from NATO's integrated military command and launched a nuclear-development programme to make France the fourth nuclear power after the USA, USSR and UK.

The question was where to test these dangerous and somewhat puzzling new weapons.

France had originally begun nuclear testing in the Sahara Desert in Algeria, but with Algeria enmeshed in a bloody struggle for independence from the late 1950s, Paris was forced to look elsewhere. With its colonial territories in the empty Pacific, and the United States having set a precedent for overseas testing in the Marshall Islands, Paris felt justified in looking to the vast Pacific to trial its new and unpredictable hydrogen bomb. Far away French Polynesia, deep in the South Pacific, seemed the perfect location.

Moruroa, an uninhabited atoll in the Tuamotu Archipelago, was chosen as the secret testing ground. (By a supreme irony, Moruroa means 'big secret' in Tahitian.) The island was large enough for an airport, distant from prying eyes in the South Pacific and at 1,610 kilometres to the east, well downwind from Tahiti and other inhabited islands, which would minimise the risk from any radioactive fallout.

France's big problem, after the Pacific islanders' experience of the US Bikini tests, was the reaction of the Polynesian people. French officials tried in vain to quell dissenting voices and political outcry. In 1958, after giving a speech in Papeete, a pro-independence local politician named Pouvanaa a Oopa, who was also known to be opposed to the tests was arrested by the French authorities on the grounds of employing inflammatory language to incite a rebellion. He was sentenced to eight years in a French jail and exiled from French Polynesia for fifteen years, being required to reside in France during those years. Despite the widespread local protests, the President of the Polynesian General Assembly, Jacques-Denis Drollet, eventually agreed to the testing. This announcement was followed by international protests, boycotting and riots. Nevertheless, the French nuclear test programme went ahead.

By 1960, the French government had begun implementing a plan both to appease the locals and to build the infrastructure for the military facilities the nuclear tests required. An all-important airport was hastily built in Faa'a in Tahiti; the French government put a positive spin on the project, claiming it was essentially a civilian infrastructure project designed to create jobs and benefit business, specifically by improving tourism. The locals were not told about the test plans until 1963 for fear that there would be protests. Paris's nervousness was well justified. When the news of the nuclear testing programme finally broke, the natives were deeply unimpressed. The first test took place at Moruroa Atoll on 2 July 1966; nearly two hundred more would be staged in the region over the next thirty years.

The exasperated New Zealand and Australian governments would eventually take France to the International Court of Justice in 1973 to try to halt the tests. France, however, would refuse to follow the court's ruling that it should stop testing. Norman Kirk, New Zealand's prime minister, even sent two Royal New Zealand Navy frigates to protest on the edge of the Mururoa testing area. France ignored them. Resistance became more organised and led to an international campaign coordinated by the environmental group Greenpeace. The latter was seen as the leader of the troublemakers by the French government and military, and with some justification: in 1985, Greenpeace piled pressure on Paris by helping to encourage the South Pacific Forum to pass the Treaty of Rarotonga, intended to create a South Pacific Nuclear Free Zone. France refused to obey this prohibition and continued testing, despite further protests. The country was firmly in the dock at the court of world opinion and didn't like it.

Events came to a head on 10 July 1985. The Greenpeace protesters' flagship *Rainbow Warrior* was moored in Auckland, New Zealand, as part of Greenpeace's plan to obstruct any French nuclear testing in the Moruroa Atoll. Most of the small crew were awake in the mess drinking beer. At midnight there were two explosions and the vessel started to list. The startled crew scrambled to safety on the wharf. Within minutes, the *Rainbow Warrior* had capsized.

'I stood there looking at the boat with all of these bubbles coming out of it,' Captain Pete Willcox would later recall. 'That's when Davey Edwards said, "Fernando is down there." I remember arguing with him, saying, "No, Fernando has gone to town." That's what he always did.

'"No," he said. "Fernando is down there."'

Fernando Pereira was the Greenpeace photographer on board to photograph the French nuclear tests. He was trapped below and drowned. Next morning, investigators found that the ship had been holed by two explosive devices, one near the propeller and one alongside the outer hull of the engine room.

Initially, the French government denied all knowledge of the sinking and joined in the condemnation of what it described as a terrorist act, but it soon became obvious that it had been involved. In an attempt to neutralise the ship ahead of its planned protest, French secret service agents in diving gear had attached two packets of plastic-wrapped explosives below the waterline. 'Opération Satanique' turned out to be a public relations disaster for the French. Eventually, an embarrassed Prime Minister Laurent Fabius appeared on French television to tell a shocked public, 'Agents of the DGSE [Secret Service] sank this boat. They acted on orders.'

The New Zealand authorities moved quickly and, after one of the country's largest police investigations, identified and arrested two of the French agents. The other members of the French team all escaped. Christine Cabon, a spy who had infiltrated the Greenpeace New Zealand office ahead of the bombing, evaded arrest and fled to Israel. She hasn't been seen since. The whereabouts of the combat frogman who bore the alias 'Jean-Michel Berthelot' – one of the two divers believed to have planted the bombs – are unknown.

In the end, only those two agents ever stood trial. Dominique Prieur and Alain Mafart, who had posed as Swiss tourists, pleaded guilty to charges of manslaughter and wilful damage, attracting sentences of ten and seven years. A UN-negotiated settlement saw them transferred to Hao Atoll, a French military base in French Polynesia. They were both quietly released within less than two years.

Throughout this period, international attempts were afoot to end nuclear testing and nuclear proliferation. As early as 1963, all the nuclear powers except China and France had signed the Nuclear Test Ban Treaty. And in 1995, when President Jacques Chirac announced plans to conduct a further eight tests, the French government was shaken by the strength of the response. New Zealand and Australia began boycotting French products and companies and, in a rare political move, both countries temporarily removed their French ambassadors. Ultimately, only six new tests were completed.

The international response to this resumption of testing was dwarfed by the outrage among the inhabitants of French Polynesia. Already poor, dissatisfied and tired of the social upheaval brought about by testing, the Polynesians erupted in protest. The streets of Papeete were shut down by massive sit-ins after the new Greenpeace protest boat, *Rainbow Warrior II*, was denied entry into the harbour. Following the first new test on 5 September 1995, rioting broke out. Part of the airport in Faa'a was burned down and the city was ravaged. Most of the looters were young, dissatisfied men. As the Mayor of Faa'a and President of French Polynesia, Oscar Temaru, pointed out, 'Is it any surprise that they turn to violence and looting . . . ?' France had badly underestimated the explosive political consequence of testing nuclear bombs on other people's property. Finally the centime dropped – nuclear testing in French Polynesia ended in 1996.

The US experience at Bikini and the French experience of the social and political consequences of nuclear testing had proved a powerful point. Nuclear weapons, even far away in the remote vastness of the Pacific, had been shown to have an unexpectedly powerful impact far greater than their potential for destruction. Politicians and nuclear scientists were now well aware that they were playing with fire – in every sense.

There is, however, one happy footnote to the miserable history of nuclear weapons in the Pacific. After the bombing, the damaged hull of the *Rainbow Warrior* was given a final resting place, scuttled and sunk in New Zealand's Cavalli Islands near Matauri Bay. It has become a living reef, attracting marine life and recreational divers.

Poor Fernando Pereira has his monument, 'full fathom five'.

CHAPTER 7

STOP DIGGING! CIVILIAN ENGINEERING

'It's hard not to be a bit schizophrenic as an engineer; we spend half our time building things and the other half learning how to blow them up!'
– MAJOR GENERAL JOHN DREWIENKIEWICZ, ENGINEER-IN-CHIEF, BRITISH ARMY

As early as 1962, engineers became interested in the potential of nuclear explosions. In a seminal paper published in 1946, a British engineer listed the uses of explosives for civil purposes. The list was surprisingly long:

- Road-making
- Excavation for railways
- Quarrying
- Secondary blasting
- Tunnelling
- Delay detonators
- Demolitions
- Mass concrete foundations

- Concrete dams
- Opencast mining

The exploration of the possible civil-engineering uses of nuclear explosives began in 1950, with substantial progress made in the ensuing years, to the extent that in 1960, a United States programme called 'Project Plowshare' under a team of scientists at the Lawrence Livermore Laboratory in California confirmed the value of using nuclear explosions for peaceful construction purposes. Successful suggestions for the non-combat use of nuclear explosives included rock blasting, the manufacture of synthetic chemical isotopes and unlocking some of the mysteries of the Earth's deep crust by probing with seismology, which would help geologists and mining companies involved in prospecting.

Project Plowshare's first suggestion was a feasibility study to see if 'Project Chariot', a planned experiment to use several nuclear detonations to excavate a harbour on the far northwest coast of Alaska, was feasible. This was never followed up, as resistance from some of the scientists, local Native American groups and a number of individuals and organizations in the continental United States forced a rethink. Instead, a programme of thirty-one nuclear warheads detonated in twenty-seven separate tests was substituted to investigate what were billed as 'peaceful nuclear explosions'.

On 6 July 1962, a 104-kiloton thermonuclear explosion, code-named 'Sedan', was detonated underground, the biggest cratering shot of the Plowshare series of tests. The device was buried 195 metres below the desert floor at Yucca Flat within the Nevada Test Site, 105 kilometres from Las Vegas. A smooth 390-metre by 98-metre crater was created by the blast. Yucca Flat was subsequently the site for 739 nuclear tests and has been called 'the most irradiated, nuclear-blasted spot on the face of the Earth'.

Project Plowshare's tests generated a storm of protest and serious public opposition. Environmental consequences included tritiated water (essentially radioactive water containing tritium, a radioactive isotope of hydrogen) that the commercial CER Geonuclear Corporation predicted would reach a level 2 per cent above the maximum permitted for drinking water. Moreover, fallout from radioactive material was ejected into the atmosphere before underground testing was finally ended by international treaty. These negative impacts led to the programme's termination in 1977.

In March 2009, *Time* magazine identified the 1970 Yucca Flat Baneberry test – during which eighty-six workers were exposed to radiation, and which carried radioactive fallout over at least four other American states – as one of the world's worst accidental nuclear disasters. After the Baneberry incident, nuclear testing at the Nevada Test Site was suspended for six months pending investigation. The US Atomic Energy Agency Commission official report concluded that the primary cause of the extensive radioactive downwind plume was 'an unexpected and unrecognised abnormally high water content in the medium surrounding the detonation point'.

Despite this, within seven months of the excavation the bottom of Sedan Crater could be safely walked upon with no protective clothing and photographs were taken. Tours of the crater are now a visitor attraction and, because the craters at the Nevada Test Site have features similar to lunar craters, eleven of the twelve American astronauts who have walked on the Moon were trained here in crater topography, geological studies and sample collection.

So, the only positive outcome of the hugely expensive underground peaceful Plowshare nuclear tests turned out to be a training ground for astronauts and a Nevada tourist attraction.

The United States continued conducting other underground

nuclear tests for more than twenty years until the final Divider test in September 1992. In 1996, the USA became the first country to sign the Comprehensive Nuclear Test Ban Treaty, which outlaws all nuclear explosions. The United States, however, is also one of the few countries that have yet to ratify the treaty for it to become global law.

Not to be outdone on the nuclear civil-engineering front, the Soviets pursued similar ideas. The Cold War wasn't only confined to military competition. Moscow decided to investigate the possibility of using nuclear power for peaceful construction purposes, in particular moving massive quantities of earth cheaply, digging new canals and reservoirs, drilling for oil and building dams. The result was a series of nuclear explosions in northeastern Kazakhstan as part of the 'Nuclear Explosions for the National Economy' project, which mirrored Operation Plowshare. Having borrowed this bad idea from the USA, the Soviet programme became much larger than Plowshare, mainly through the number of nuclear explosions – 156 to the USA's 27. The Soviet programme also lasted much longer, only being wound up in 1989, a decade after the final Plowshare tests.

Perhaps the most spectacular test was in Kazakhstan at Chagan, a virgin forest site on the border of the Semipalatinsk Test Site, in January 1965. The Chagan test was designed to investigate the suitability of nuclear explosions for creating reservoirs. A 140-kiloton device was placed in a 178-metre deep hole in the dry bed of the Chagan River, a tributary of the Irtysh. This was the first and largest of all detonations carried out during the Kremlin's 'Nuclear Explosions for the National Economy' programme. The aim was to build a crater lip that would dam the river during periods of high flow.

The test went badly wrong, however, because the explosion was much larger than the scientists had predicted and the blast hurled a huge radioactive plume into the atmosphere, spreading contamination as far as Japan and the Far East. It drew a furious protest from Washington, which accused the Soviets of breaching the 1963 Limited Test Ban Treaty, which outlawed all atmospheric tests. The Soviets shamefacedly replied that it had been an underground test and the quantity of radioactive debris that escaped into the atmosphere was insignificant. The first statement was true; the second was not. Scientists now knew that all these underground tests really achieved was to blow radioactive dust into the stratosphere to be carried halfway round the world.

Yet at first the results on the ground seemed to be a great success for Moscow. The underground explosion had blasted a huge hole and created a crater some 400 metres across and 100 metres deep with a lip height of 20 metres to 38 metres. Soviet engineers then dug a water channel into the crater to fill a new lake and to create a reservoir behind the crater lip. Spring melt soon filled the crater with 6.4 million cubic metres of water, and the reservoir behind it was filled with a further 10 million cubic metres.

At that time of its creation, the Soviet government was delighted with what was seen as the success of Lake Chagan. They even made a public-relations film in which the minister responsible for the entire Soviet nuclear weapons programme was seen taking a swim in the crater lake and drinking its water. Water from the new lake was given to cattle in the area.

The scientists then discovered a problem. The lake water was actually highly radioactive. Not only that, the channel used to control excessive overflow was pumping radioactive water into the nearby Chukchi Sea, which lies between Russia and Alaska to the north of the Bering Strait.

The reservoir, known informally as Lake Chagan, still exists today. Radioactivity 100 metres away from the lake is at background level, but the lake water is still contaminated and definitely undrinkable. There are no fish, no wildlife, no birds and a disgusting stink often comes off the water. Swimming is not permitted, for obvious reasons, and there is evidence that the contaminated water is still draining into the Irtysh River.

Efrim P Slavskiy, Minister of the All Soviet Medium Machine Building Ministry (the department that oversaw the Soviet nuclear weapons programme) was the lucky person to have taken that first swim in the nuclear waters of the Chagan crater lake. He apparently lived on and worked for many years, to be persuaded to resign at the age of eighty-eight in the wake of the Chernobyl disaster.

He was, allegedly, recorded as saying that the water had tasted disgusting, and then calling for vodka...

CHAPTER 8

THE NUCLEAR NAVY

'Good ideas and innovations must be driven into existence by patience and courage.'
– ADMIRAL HYMAN RICKOVER, US NAVY

The idea of a submarine is much older than most of us think. Just as mankind has always dreamed of flying so, since classical times, humanity has also pondered the possibilities of underwater travel – usually for the purpose of getting at their enemies unseen. The first recorded attempt of a submarine attack comes from Classical Greece in about 413 BC, when unseen divers breathing through hollow reeds cleared obstructions at the Athenian siege of Syracuse, according to Thucydides' *History of the Peloponnesian War*.

If anyone deserves the credit for the first workable submarine, it is Cornelis Drebbel, a Dutch doctor living in England, who came up with what might be called the first 'practical' submarine in 1620. This was a boat with a lead keel that sank to just below the surface, powered by rowers pulling on oars that protruded through flexible leather seals in the hull. Snorkel air tubes floated on the surface. Drebbel's submarine is recorded as successfully manoeuvring at depths of around four metres below the surface.

Despite successful demonstrations in London's River Thames, Drebbel's invention failed to arouse the interest of the Royal Navy. Not unreasonably, they could not see its usefulness in the rougher waves of the open sea. But Drebbel's dream did not die. For the next two hundred years, naval architects and inventors struggled to come up with a vessel that could operate successfully underwater, plant a weapon and keep the crew alive.

The big problem was propulsion and in 1865 German-American engineer Julius H Kroehl designed and built his *Sub Marine Explorer*, the first submarine to successfully dive, move and emerge to the surface under control. His design incorporated many ideas that are still used by modern submarines, such as blowing ballast by compressed air. After its public maiden dive in 1866, the craft went on to work collecting pearls off Panama. The vessel could dive to depths of thirty metres and then surface quickly – rather too quickly, in fact, which resulted its crew and pearl fishers developing 'the bends', or decompression sickness, a condition unknown at the time. But it can still claim to be the first real working submarine.

War breeds inventions. After the American Civil War, a blizzard of new designs, patents, and hasty ship-building finally culminated in the Holland boats of 1900. Irish inventor John Philip Holland built a working model as early as 1876 and a full-scale version two years later. In 1897, after falling out with his Fenian Irish-rebel backers, he launched the privately funded 'Holland Type VI' submarine. This was the first craft able to run submerged for any considerable distance, using battery-powered electric motors charged by its internal combustion engines when running on the surface. An initially sceptical, then increasingly interested, US Navy bought the boat in 1900. Following rigorous tests she was commissioned on 12 October

1900 as USS *Holland*. It was a breakthrough. In 1902, Holland received US Patent 708,553 for his modern submarine and many other nations suddenly took an interest in the offensive potential of this new weapon, buying up the rights to build under licence – notably the German Kaiserliche Marine and Britain's Royal Navy. The age of the submarine had arrived.

World War I proved the submarine's deadly effectiveness. In April 1915, the German *U-20* torpedoed the liner RMS *Lusitania*, sailing alone and without protection off the south coast of southern Ireland. The ship sank in eighteen minutes, killing 1,198 people, with 761 survivors. The incident signalled two things: the deadly tactical effectiveness of the submarine at sea and its ability to send a potent political message. A huge international row erupted as many of the *U-20*'s victims had been neutral Americans. Of the 139 US citizens on board *Lusitania*, 128 lost their lives. The attack caused massive outrage in America as well as in Britain. Leaving the tactics and wider political reverberations to one side, World War I made it abundantly clear that submarines worked.

After the German fleet's flight for the safety of Kiel after the battle of Jutland in mid-1916, the Kaiser's only real riposte to the might of the Royal Navy was unrestricted submarine warfare. This turned out to be spectacularly successful and by spring 1917 had brought Great Britain to within six weeks of running out of food. Only the convoy system and improved tactics staved off defeat.

The two world wars taught several lessons about submarines, not least that they were a strategic weapon. They could strangle an enemy's trade, but as long as they had to surface regularly to recharge their big, heavy electric batteries and clear the air, they were vulnerable on the surface. If an engine could be developed that allowed a vessel to stay below the waves, making its own oxygen and operating without surface refuelling, then a submarine would become effectively unassailable.

The US Navy's search for such a propulsion system led to a grouchy workaholic called Hyman Rickover. Rickover had been born in what was then part of Tsarist Russia and his family fled to America in 1906 to escape the anti-Semitic pogroms sweeping Poland after the failed Russian Revolution of 1905. The family eventually settled in a Jewish neighbourhood in Chicago and young Rickover took his first paid job at age nine. The family became friendly with a congressman who was also a Jewish immigrant. He nominated Rickover for appointment to the United States Naval Academy at Annapolis, from which he graduated in 1918.

Rickover served on the *s-9* and *s-48* submersibles for four years from 1929, and subsequently translated the German Imperial Navy's *Das Unterseeboot* ('The Submarine'), which became the basic text book for the US submarine service. During and after World War II, and despite his undistinguished war record ashore, Rickover began to make a name for himself as the US Navy's star electrical engineering officer. After the war, the United States was looking for secondary applications of the mouth-watering power offered by the new atomic energy from nuclear reactors. Rickover had also realised its potential and began to lobby for a nuclear-powered navy. By 1948, he had become fixated on the possibilities of nuclear-powered submarines.

Rickover knew that atomic energy would revolutionise the submarine, making it a true underwater vessel rather than a submersible craft that could only stay underwater for limited periods. Such a submarine would have the ability to operate submerged at high speeds, comparable to those of surface vessels, for unlimited periods, dependent only on the endurance of its crew. Theoretically, the crew would starve before the submarine ran out of fuel or oxygen.

Initially, he faced resistance from his superiors and was consigned to 'advisory duties'; his office was a former ladies' lavatory. Frustrated, in 1947 he made a direct appeal to the Chief of Naval Operations, Admiral Chester Nimitz, himself an ex-submariner. He understood immediately what his pushy subordinate was advocating, overrode the navy's naysayers and recommended the nuclear project to his superior, John L Sullivan – the Secretary of the Navy. Despite political skulduggery by Rickover's jealous rivals, Sullivan ordered the navy to proceed with building a nuclear-powered vessel. Rickover himself later credited Secretary Sullivan as being 'the true father of the nuclear navy'.

In July 1951, the United States Congress authorised the construction of the world's first nuclear-powered submarine for the US Navy, the USS *Nautilus*. General Dynamics in Groton, Connecticut, began the shipbuilding process in June 1952. Much larger than the diesel-electric submarines that preceded it, the new submarine stretched 97 metres and displaced 3,230 tonnes.

The challenge of developing a suitable small reactor was given to the Westinghouse Electric Corporation. The reactor had to be compact enough to fit into a narrow submarine, but it had to be safe as well as small. Static, and on shore where there is lots of space, this is not too difficult, but at sea a nuclear reactor has to be able to withstand the stresses of operating in the real, tightly enclosed and sometimes dangerous world of maritime warfare, where ships move about.

Nuclear reactors are basically simple steam engines: a kettle, heated by a controlled nuclear reaction. As uranium fissions and the atoms split, the energy comes in the form of heat, which can then be used to produce pressurised steam. That drives the turbines that provide the submarine's actual power. By 1954, Westinghouse had developed a Submarine Thermal Reactor (STR) designed to produce 13,000 horsepower. The STR worked by using fuel rods made of the highly enriched isotope uranium-235, to maximise the amount of nuclear fuel in the core, helping to create a more compact reactor and boost longer core life. It was cooled by water and was technically a small pressurised water reactor (PWR).

USS *Nautilus* was launched on 21 January 1954 by Mamie Eisenhower and commissioned on 30 September 1954. The STR reactor on the *Nautilus* was started on 17 January 1955 and propelled the sub's maiden voyage, making history for being first 'underway on nuclear power'. In 1958, it became the first vessel to reach the geographic North Pole. Shortly afterwards, the first nuclear-powered aircraft carrier also had her maiden voyage. The USS *Enterprise* was launched in September 1960 and commissioned on 25 November 1961. The 'Big E's' nuclear reactors were finally decommissioned in 2017 after nearly sixty years' service.

Hyman Rickover's dream of a nuclear navy had become reality. Sharp-tongued, intolerant and scathing of his superiors, Rickover drove his men hard and was equally uncompromising with contractors – especially those he caught milking US tax dollars. *Time* magazine featured him on the cover of its 11 January 1954 issue as 'a man *who gets things done*'.

Remarkably, Rickover served in a flag rank for almost thirty years (1953 to 1982), retiring as a four-star admiral at the age of eighty-two. With sixty-three years of active duty service, he remains the longest-serving member of the US armed forces.

Rickover's stringent standards are largely credited with being responsible for the US Navy's continuing record of zero reactor accidents. The mishap-free record of United States Navy reactor operations stands in stark contrast to those of the Soviet Union.

Although two US nuclear submarines have sunk over the years, there is no evidence that any of the power plants Rickover engineered in his long tenure ever broke down or endangered anyone, a record the civilian nuclear power industry has not been able to match. As David Hitzfelder, an engineer working on US Navy nuclear submarines noted in 2017:

There are many dangers involved in the operation of a navy nuclear reactor, that range from burns from touching hot equipment, electrical shock from coming into contact with electrical control components, potential steam leaks killing the crew, increased risk of cancer due to radiation exposure while working in the reactor compartment, and a host of other potential dangers.

The US Navy has systems and procedures in place to minimise the risks associated with operating a navy nuclear power plant. As long as the crew follows procedure, and the equipment is properly maintained, there is very little risk of serious accidents. Radiation exposure is monitored, and the reactor compartment is heavily shielded from radiation, and locked during operation to minimise the health risks from radiation exposure. The reactor itself is very stable, and has safety engineering features to prevent nuclear accidents, or minimise the potential damage and radiation leakage should an accident occur. For that we have to thank Rickover...

Of the two US Navy nuclear submarines lost, neither – as far as we know – was sunk by reactor problems. The first such incident was the disappearance and sinking of the USS *Thresher* on 10 April 1963, with 129 sailors on board. This remains the US Navy's highest submarine death toll. The reason for its loss remains a mystery.

The Threshers were designed to be fast, deep-diving nuclear attack submarines. Their pressurised water reactor drove two steam turbines. The lead boat in her class, *Thresher* sank while conducting a dive to her test depth of about 1,100 feet (335 metres). Five minutes before losing contact with the boat, the submarine rescue ship *Skylark* received a garbled underwater transmission that *Thresher* was having some minor technical problems.

Skylark reported further garbled messages until the sonar picked up the sound of *Thresher* imploding at between 1,300 and 2,000 feet (400 and 600 metres).

A Naval Court of Inquiry concluded the probable cause of the *Thresher*'s loss was 'major flooding' caused by a piping failure, a finding that has since been challenged by naval and submarine experts. After more than half a century, eighteen pages of testimony from key witnesses still remains closed to the public. The inquiry justified its finding by citing a then recent history of silver-brazed pipe joint failures on six submarines, including the *Thresher*. Many suspected that something had gone wrong with the nuclear reactor but, without evidence, this could not be cited as the cause of the sinking. Had the reactor been in an emergency shutdown, or reactor 'scram', as the emergency was declared? (A scram is the rapid insertion of the control rods into the core to stop the fission chain reaction, a key safety function of a nuclear reactor.) If so, why?

The waters were further muddied by a 1987 interview with Fred Korth, who was Secretary of the Navy when the *Thresher* was lost, and his executive assistant, Vice Admiral Marmaduke Bayne. Both stated that Admiral Rickover had personally altered portions of the Naval Court of Inquiry's report, and had probably done so to ensure that the report's findings were left as 'inconclusive'. This finding deflected blame for the sinking away from Rickover's beloved naval nuclear reactors by creating doubt that there had been an emergency scram. We will never know.

The second US nuclear submarine loss was more clear-cut and cannot be attributed to any nuclear problem. The USS *Scorpion* disappeared off the Azores in May 1968 with ninety-nine sailors on board. Curiously, three other submarines disappeared in the same year, all without explanation. Russia's *K-129*, France's *Minerve* and Israel's INS *Dakar* were the others.

Precisely what happened to the *Scorpion* is still a mystery –

the boat simply failed to return to port on 27 May. On 21 May, it indicated its position to be about eighty kilometres south of the Azores, where it was gathering intelligence on a Soviet task group. Six days later, it was reported overdue at Norfolk. A search was initiated but, on 5 June, *Scorpion* and all hands were declared 'presumed lost'. The navy had launched a sudden search three days before the disappearance was reported, which indicates that they knew something was up. Eventually, *Scorpion* was located in 3,050 metres of water by a navy research ship. US Navy sources suggest that 'the most likely cause was an inadvertent activation of the battery of a ... Mark 37 torpedo ...'

Subsequently, the Court of Inquiry was reconvened and other vessels, including the submersible *Trieste II*, were dispatched to the scene, but, despite myriad data and pictures collected and studied, the cause of the loss remains a mystery. One thing appears clear from the acoustic records and the photographs of the wreck: failure of the nuclear reactor was not the cause.

Despite the risks, the US Navy – with some justification – can be proud of its nuclear safety record at sea.

CHAPTER 9

ALL AT SEA: RUSSIA'S TERRIBLE RECORD

'There is nothing more enticing, disenchanting, and enslaving than the life at sea.'
– JOSEPH CONRAD

Unlike its US counterpart, the Russian Navy has an appalling nuclear safety record. Worst of all was the loss of the giant *Kursk*, which is afforded its own detailed chapter later on. *Kursk*, however, was just one of many such incidents. The full list of thirty years of Soviet nuclear submarine accidents, compiled in 1996 by the Risar National Laboratory at the Danish Technical University, makes for horrifying reading:

September 1960: Soviet November-class submarine (*K-8*) – accident.
July 1961: Soviet Hotel-class submarine (*K-19*) – accident.
1964: Soviet nuclear submarine (*K-27*) – incident.
February 1965: Soviet November-class submarine (*K-11*) – accident.
November 1965: Soviet Echo I-class submarine (*K-74*) – accident.

Autumn 1966: Icebreaker NS *Lenin* – accident.
1966: Soviet nuclear submarine – accident.
September 1967: Soviet November-class submarine (*K-3*) – accident.
1967: Soviet November-class submarine – incident.
May 1968: Soviet nuclear submarine (*K-27*) – accident.
August 1968: Soviet Yankee-class submarine (*K-140*) – accident.
1968: Soviet nuclear submarine – accident?
April 1970: Soviet November-class submarine (*K-8*) – accident.
1970: Soviet Charlie II-class submarine (*K-320*) – accident.
April 1971: Soviet Yankee-class submarine – incident/accident?
1971: Soviet submarine radiation – incident/accident.
February 1972: Soviet Hotel II-class submarine (*K-19*) – accident.
March 1972: Soviet Yankee-class submarine – accident?
December 1972 – January 1973: Soviet Yankee-class submarine – accident.
1972: Soviet Alfa-class submarine (*K-377*) – accident.
June 1973: Soviet Echo II-class submarine (*K-56*) – accident.
August 1973: Soviet Yankee-class submarine (*K-219*) – accident.
April 1974: Soviet Yankee-class submarine (*K-420*) – incident.
March 1976: Soviet nuclear submarine – accident.
September 1976: Soviet Echo II-class submarine (*K-47*) – accident.
October 1976: Soviet Yankee-class submarine – accident.
1977: Soviet Echo-II-class submarine – accident.
1977: Soviet Nuclear submarine radiation – accident.
August 1978: Soviet Echo II-class submarine – incident.
September 1978: Soviet Yankee-class submarine (*K-451*) – accident.

December 1978: Soviet Delta I-class submarine (*K-171*) – accident.
July 1979: Soviet Echo I-class submarine (*K-116*) – accident.
December 1979: Soviet nuclear submarine – accident.
August 1980: Soviet Echo II-class submarine – accident.
September 1980: Unconfirmed Soviet nuclear(?) submarine – accident.
September 1980: Soviet nuclear submarine (*K-222*) – accident.
1980: Soviet Delta III-class submarine – accident.
April 1982: Soviet Alfa-class submarine (*K-123*) – accident.
June 1983: Soviet Charlie I-class submarine (*K-429*) – accident.
September 1983: Soviet nuclear submarine – accident.
June 1984: Soviet Echo I-class submarine (*K-131*) – accident.
August 1985: Soviet Echo II-class submarine – accident.
December 1985: Soviet Echo II/Charlie-class submarine – accident.
1985: Soviet Echo II-class submarine – accident.
January 1986: Soviet Echo II-class submarine – incident.
October 1986: Soviet Yankee-class submarine (*K-219*) – accident.
April 1989: Soviet Mike-class submarine (*K-278*) – accident.
June 1989: Soviet Echo II-class submarine (*K-192*) – accident.
July 1989: Soviet Alfa-class submarine – incident/accident.
December 1989: Soviet Delta IV-class submarine – accident.
January 1990: Soviet Admiral Ushakov-class cruiser *Kirov* – accident.
October 1991: Soviet Typhoon-class submarine – accident.

The difference in standards between the Soviet approach to nuclear safety at sea and that of the US Navy can perhaps best be illustrated by the story of the fire aboard Russia's first nuclear-powered attack submarine – *K-3 Leninsky Komsomol* – which was nearly destroyed in September 1967 in a fire caused by a jury-rigged repair. Long afterwards, a former crew member told *Pravda*, 'A non-standard gasket from . . . a beer bottle, I think, was installed in the ballast tank.' He added, 'Naturally it was displaced, the hydraulic fluid leaked under the pressure of 100 atmospheres and got sprayed on to the lamp, which had a broken protective cap. Inflammation occurred immediately. We had a panic to put it out.'

With the beer-bottle approach to nuclear engineering, it is hardly surprising that Soviet nuclear safety at sea has left much to be desired. The first major Russian nuclear submarine disaster was the ill-fated *K-19*, wryly nicknamed 'Hiroshima' by its worried crew. Sadly, *K-19* was merely the first in the long list of Russian nuclear submarine disasters of the past several decades. On the vessel's very first trial sailing, the reactor broke down.

In many ways the first was the worst. The indefatigable David Lowet of Stanford University studied the story of the ill-fated submarine in great detail. His findings do not make for happy nuclear reading. The sudden success of the USS *Nautilus* had shocked the Soviet Politburo. Soviet leaders were determined to build a nuclear submarine fleet that would rival that of the United States. The keel of the first nuclear boat was laid down as early as September 1958. Right from the start, *K-19* was an unlucky ship. In 1959, an accidental explosion killed two workers. Soon after, six women died, overcome by fumes while gluing rubber lining into a water cistern. For sailors, ever superstitious, the most persuasive omen that *K-19* was a cursed boat came at the launching, when a champagne bottle struck across the bow bounced off, unbroken.

When the boat was launched and named *K-19*, it was obvious, even to the new captain, Nikolai Zateyev, that this was a dangerously rushed programme. He even warned fleet headquarters that his new command was not fit for combat. He was right: *K-19* turned out to be a bird of ill omen.

The submarine was completed in November 1960 and had very limited capabilities. It could only operate at a maximum depth of 500 feet (150 metres) with just three ballistic nuclear missiles on the submarine, each with a range of 650 kilometres. Production and testing was so rushed that *K-19*'s sea trials were plagued by breakdowns and malfunctions. However, in June 1961, the sub finally departed into the North Atlantic on its first operational mission.

Things soon went wrong. On 4 July 1961, *K-19* developed a radioactive leak deep in the Atlantic south of Greenland. The leak was caused when the first reactor pressure test went all the way to 400 atmospheres because of a pressure gauge malfunction. The designed pressure was only 200 atmospheres, so this resulted in damage to the piping of the primary system. In true Soviet style, this incident was not reported up the chain of command and the captain remained unaware of the problem, so the necessary repair work was not performed. Unfortunately, the leak was located in a pressurised pipe within the primary cooling circuit, causing a sudden drop in pressure, which set off the reactor emergency systems. This drop in pressure led to the reactor water boiling. The reactor room became too hot and a fire broke out. The fire was extinguished, but there was still a major problem: how to cool the runaway reactor core and stop a nuclear meltdown. With no coolant system in place to stop the reactor from overheating, the crew, protected only by raincoats and gas masks, had to enter the reactor compartment and fix the leak, in an effort to save the submarine from exploding. One of the first volunteers was a popular young officer, Lieutenant Korchilov.

'I accompanied him to the reactor room door – to his death,' Captain Zateyev later remembered, 'and I said, "Well, Boris, do you know where you're going?"

'And he said, "I know, Comrade Captain."'

Five minutes later, Korchilov stumbled out of the reactor room, tore off his gas mask and vomited, but his team's heroic efforts saved the boat. They solved the problem by using the cool water in the crew's drinking water tank. They had kept the reactor from exploding, but the damage-control team had exposed themselves to massive overdoses of radiation from the contaminated gas and steam.

Captain Zateyev recalled, 'Right on the spot, their appearances began changing. Skin not protected by clothing began to redden, face and hands began to swell. Dots of blood began to appear on their foreheads, under their hair. Within two hours we couldn't recognise them. People died fully conscious, in terrible pain. They couldn't speak, but they could whisper. They begged us to kill them.'

The crew was evacuated by another Soviet submarine and *K-19* was towed home to base on the Kola Peninsula. Korchilov and five other sailors were rushed to Moscow. Within ten days, all six were dead. Their radioactive corpses had to be buried deep in lead coffins.

Several of the other compartments on the submarine, and the rest of the crew, also became contaminated. In total, twenty-three crew members died of radiation poisoning – eight in a matter of days, the rest over the next two years. The submarine was later repaired by replacing the reactor with a new nuclear power unit, and brought back into service.

The submarine's woes continued. There was another reactor accident on *K-19* in 1972. The fate of the *K-19* remained a closely guarded secret, and was not publicised to Western sources until 1991, when the newspaper *Pravda* confirmed that radiation had killed many members of the crew. The crew members were even ordered to lie to doctors in routine check-ups decades after the incident.

K-19 was far from an isolated case. The keel of *K-27* was laid down as early as 15 June 1958, but it wasn't until October 1963 that she entered service, as an experimental attack submarine, joining the Soviet Northern Fleet two years later. As her construction had proved so expensive, and had taken far longer than other Soviet nuclear submarines, sailors dubbed her '*Zolotaya Rybka*' ('Little Golden Fish') after a story, best known from a version by Alexander Pushkin, in which a glittering captured fish promises to grant any wish in exchange for its freedom.

K-27 was unusual in carrying two experimental, liquid metal-cooled reactors – a design never tried before in the Soviet Navy. The nuclear reactors were troublesome from their first criticality, but at least the boat was able to engage in operations. Vyacheslav Mazurenko, then aged twenty-two, was serving as a chief warrant officer on the vessel. In 2013, he told the BBC Russian service what happened: 'In Soviet times, we were told that our subs were the best, and we had to be different from the imperialists.' The crew were regarded as part of an elite military elite . . . and were accorded special privileges, including foodstuffs denied to ordinary Russian people. 'But the first subs were far from perfect,' Mazurenko admitted. 'Soviet leader Nikita Khrushchev said, "We'll catch up with you and overtake you." They kept churning out new subs, regardless of the risk to people . . . When the assessment commission came round, its members were often afraid to visit the reactor compartment. They always tried to avoid it, but Captain Leonov actually sat on one of the reactors, to show them how safe it was.'

At the height of the Cold War, however, *K-27* met with disaster when something went badly wrong with her reactors during a voyage in the Arctic. On 24 May 1968, she suffered an abrupt drop-off in power from one reactor. Her engine room filled with radioactive gas; radioactivity levels – largely from thermal neutrons and deadly gamma rays – inside the vessel soared. According to Mazurenko, 'We were on a five-day trip to check everything was working normally, before a seventy-day round-the-world mission without resurfacing.

'It was the end of the third day and everything seemed to be going well. The crew was really tired. The bulkheads were open. I was in the fifth compartment, next to the fourth compartment with the two nuclear reactors. We suddenly noticed some people running.

'We had a radiation detector in the compartment, but it was switched off... But then, our radiation supervisor switched on the detector in the compartment and it went off the scale. He looked surprised and worried.' As the lethal gas was odourless, it took some time for the crew to appreciate what had happened. A couple of hours later, some sailors emerged from the fourth compartment; a few were so weak that they had to be carried.

Unfortunately, owing to insufficient preparation, K-27's crew weren't aware that there had been widespread failures in the fuel elements within the reactor. The catalogue of disasters included insufficient cooling over some 20 per cent of the core of the reactor, which had also ruptured, unleashing dangerous nuclear fuel and fission substances. Eventually all attempts at repairs were abandoned, by which time nine sailors had already absorbed fatal levels of radioactivity.

According to a BBC News report of 24 January 2013, the submarine headed back to its base on the Kola Peninsula, by the Barents Sea. As the sub approached, the base's command fled the dockside, because special radiation alarms onshore were emitting a deafening roar.

Soon after, the base commander picked up the captain in a car, but most of the crew had to walk two kilometres back to their barracks under their own steam. No one would come near them. Several specialist crew members were left on board the toxic sub for about a day, because they were under orders to keep watch. Some blamed *K-27*'s captain, Pavel Leonov, over the accident, but Mazurenko says the submarine's commander faced a life-or-death choice.

'When the sub surfaced to make the trip back to the docks,' he explained, 'the division ordered it to cut its engines and await special instructions. The captain, however, decided to keep going, because if the sub stopped for several hours nobody would survive long enough to get it back to base.' Although nine members of the crew died in the immediate aftermath of the disaster, the entire crew of 144 had been poisoned during the incident, many of whom were to endure ill health for years before dying prematurely.

The Soviets' nuclear difficulties were not just confined to what went on inside their reactors; they suffered further grief from what came out as well. *K-159* is a prime example of Soviet trouble with their reactors at sea. She was launched in June 1963, but was beset by an accident on 2 March 1965 when her steam generators became contaminated with radioactivity, most likely brought about by leaks of primary coolant. If this was indeed the case, makeshift repairs must have been made to block the leaking tubes, as the submarine was in service for another couple of years. It was only in 1967 that she was overhauled, during which replacement steam generators were fitted. The problem of leaky reactor discharge was never really solved and the highly contaminated *K-159* was finally scuttled – in shallow water! – in the Barents Sea.

The Soviet Navy's nuclear problems had two main causes: leaks from badly designed – and even more badly engineered – small nuclear reactors; and, more dangerously, reactors going critical. It is of interest to note that criticality accidents have never occurred during normal reactor operation, but always when the reactors were being shut down, in connection with refuelling, changes in, or installations of, the control system, or tests of the reactor system.

Known incidents involving Soviet early criticality reactor accidents include:

- February 1965: Soviet November-class submarine (*K-11*) – accident.
- August 1968: Soviet Yankee-class submarine (*K-140*) – accident.
- 1970: Soviet Charlie II-class submarine (*K-320*) – accident
- November 1980: Soviet nuclear submarine (*K-222*) – accident.
- August 1985: Soviet Echo II-class submarine (*K-431*) – accident.

During this period, no Western nuclear vessel is known ever to have suffered a so-called 'loss-of-coolant accident' or nuclear-reactor mishap.

On 13 October 1960, one of *K-8*'s steam-generator tubes broke, which brought about a loss of coolant. Sailors managed to rig up a makeshift system to supply the reactor with water coolant, but as they worked the submarine became contaminated with off-the-scale levels of radioactive gas; many crew members received 200 rads of radioactivity, while exposure to the gas left three of them with visible injuries. Worse was to come for the unfortunate crew of *K-8*. On 8 April 1970, when it was taking part in the major 'Ocean-70' naval exercise, it experienced fires in two compartments at the same time; the conflagration passed rapidly through the vessel's air-conditioning system; its two nuclear reactors lost coolant and both had to be closed down. The contaminated sub wallowed on the surface without power and with a rising sea. The captain's initial order to abandon ship was overruled after the arrival of a salvage ship. The stricken sub was taken under tow but foundered and sank when a storm blew up as she crossed the Bay of Biscay. *K-8* went down on 11 April 1970, around 481 kilometres (260 nautical miles) to the northwest of Spain; it sank to some 4,680 metres, along with its quartet of nuclear torpedoes. This was the first reported loss of a Soviet nuclear-powered submarine.

Worse was to follow. In October 1986, the Yankee-class ballistic missile submarine *K-219* was operating north of Bermuda when seawater leaked into missile silo six (which had been sloppily welded shut ashore, following an earlier accident) and reacted with missile fuel, causing an explosive mix of gases. Weapons Officer Alexander Petrachkov opened the hatch cover, so that the missile tube could be offloaded into the surrounding water, but not long afterwards the missile silo exploded.

The explosion breached the hull and water began pouring into the submarine, which now plunged from its original depth of 40 metres to more than 300 metres. It was only saved by the rapid sealing of all of the compartments; the water was evacuated by pumps, working at maximum power; the vessel stabilised. Only then did the captain, Igor Britanov, realise that the reactor, which should have shut down automatically, was still running and not responding to controls. Sergei Preminin, a young enlisted sailor, volunteered to go into the contaminated compartment and manually close down the reactor. Working with a full-face gas mask, he managed to do so, but by then a major fire had developed in the compartment, increasing the pressure there, which in turn made it impossible for him to open the door and escape; he was asphyxiated. His heroic efforts saw Preminin awarded the Order of the Red Star posthumously.

With disaster temporarily averted, Britanov used battery power to raise his vessel to the surface. He was told to have the stricken sub towed some 7,000 kilometres to Gadzhiyevo, her home port, but this proved impossible to achieve; moreover, by now noxious gases were leaking into some of *K-219*'s compartments. Against orders, Britanov ordered the crew to evacuate to the towing ship, though he stayed on board. The furious HQ Northern Fleet command at Severomorsk ordered *K-219*'s security officer to take over command and arrange for survivors to return to the sub. Events overtook them, however: the vessel flooded and *K-219* sank some 6,000 metres into Hatteras Abyssal Plain on 6 October 1986. Britanov had abandoned ship before *K-219* made her final descent, taking her nuclear weaponry to the sea floor. Captain Britanov, who had done everything that a dutiful captain could do and shown grim devotion to duty, was charged with a number of crimes, including – astonishingly – treason. In mid-1987, however, all charges were quietly dropped.

The core issue was that Soviet standards concerning naval training, nuclear engineering and nuclear safety were simply not up to the job. If there was any doubt, the *K-278 Komsomolets* disaster of April 1989 confirmed it. The unfortunate vessel sank on its very first operational patrol.

The *K-278 Komsomolets* was launched in May 1983, as part of the Soviet Northern Fleet. She was a large boat, 117.5 metres (385.5 feet) long, with a submerged displacement of 8,130 tonnes. *Komsomolets* had two water-cooled nuclear reactors. Her inner pressure hull was made of a new titanium alloy, light and strong, making her the world's deepest-diving submarine. Her operating depth, below 915 metres (3,000 feet), was far beyond that of the best US subs. She could carry a mix of torpedoes and cruise missiles with conventional or nuclear warheads. NATO dubbed her class 'Mike' and expected this dangerous new weapon to be first of a group of large, aggressive attack submarines. She became operational in late 1984, but no further Mikes were built.

The CIA took a particularly close interest in this revolutionary Soviet submarine. It represented a serious threat to NATO subs. The Langley report revealed the full story of its demise. On 7 April 1989, *K-278 Komsomolets* was cruising at 380 metres (1,250 feet) below the surface of the Norwegian Sea, 320 kilometres north of the Norwegian mainland. The problem started when a high-pressure air line connecting to the main ballast tanks (allowing the submarine to control its depth) burst its seal in the seventh compartment. A spray of oil began a flash fire in the high-pressure, oxygen-rich air.

Chief Engineer Valentin Babenko and the commanding officer, Captain First Rank Yevgeniy Vanin, were in the control room. Babenko recommended that Captain Vanin smother the apparent fire with freon, a non-flammable gas. The skipper hesitated, however, because he knew that the freon would asphyxiate the men in the compartment. But the fire was now beyond containment.

Pressure forced burning oil into compartment six, and the fire spread through bulkhead cable runs, despite closed hatches. Turbine generators wound down automatically, the emergency system to protect the nuclear reactors from overload kicked in, and the propeller shaft stopped turning. Fearing a meltdown, the reactor officer shut down the submarine's main source of power.

Now *Komsomolets* was drifting at a depth of 152 metres (500 feet), without power. Having lost vital lift and manoeuvrability, Captain Vanin ordered all ballast tanks to be blown and managed to bring the sub to the surface. As she foundered, he signalled an encoded SOS to his headquarters.

But surfacing had not put *Komsomolets* out of danger. She was still on fire. The flames had spread through cableways to all aft compartments and temperatures had reached more than 1,000°C. The rubber coating on the outer hull, designed to muffle acoustic detection, began to slide off the hot metal. Vanin ordered all hands not engaged in damage control topside. Those fighting to save the ship donned masks, using the emergency breathing system.

The situation was now out of control. Captain Vanin abandoned Soviet security protocol and sent a Mayday message on the international hailing frequency, giving the submarine's name, location and dire circumstances. Soviet Fleet HQ responded and ordered all steps to rescue the crew, including assistance from Norway, although the Norwegians had quickly become aware of the crisis through intercepted communications.

At sea, the crew assembled on the weather deck. At 2:40 pm, a rescue aircraft broke through the clouds and spotted *Komsomolets* dead in the water, smoke pouring from her hatches. Visibility was fair, and the sea state moderate, so the crew did not don wetsuits, although the water was only a couple of degrees above freezing, cold enough to kill them in fifteen minutes. In a short time, the wind began to pick up, creating 1.5-metre (4-foot) waves, and the men had to hang on to the slippery deck. For the next two hours, things seemed to be under control. The crew cleared below and the ship was not taking on water. Surface rescue was expected to arrive at 6:00 pm. Most of the crew went back topside, because the smoke inside the ship was becoming intolerable.

Captain Vanin attempted to right his ship by counter flooding. He had been handicapped by damaged equipment, hazardous conditions and a nearly complete lack of information from his instruments. About 4:30 pm, he ordered two port ballast tanks blown to trim the sub. But the empty tanks were admitting seawater from below. *K-278 Komsomolets* began taking on water astern. As the stern settled at 4:42 pm, Captain Vanin ordered the crew to abandon ship and minutes later sent his last radio message.

At 5:00 pm, two life rafts were inflated on the bow, and the aircraft dropped a rescue pod. Men began to enter the rafts. The captain went below to get the last of his crew, but now *K-278* was sinking fast. The submarine was equipped with an escape capsule, and the survivors thought they could use that.

Topside, however, things had gone badly wrong. One life raft overturned and men crowded aboard, some clinging to the sides. The second raft went down with the sub, broke free, but drifted too far for the men to reach. More small rafts were dropped from the rescue aircraft, but there were not enough for the fifty men in the water. Inside the sinking *Komsomolets*, six men were still alive. Captain Vanin joined them in their last hope: the escape capsule. As the submarine sank, an explosion rocked the ship, and suddenly the escape capsule was thrown free to the surface. Once there, the hatch blew off. Shortly after 6:00 pm, a fishing boat arrived and picked up 30 crewmen. Of the 69 crewmembers, 39 were already dead.

In the era of 'Glasnost' the incident could not be covered up, even by the secretive Soviet media. Moreover, the Norwegians observed the rescue attempts and were worried about the release of radioactivity. Oslo claimed they could have reached the scene by air and surface two hours before the submarine sank – if they'd been asked from the start.

In all, 42 crew members eventually died, 34 of them from hypothermia in the frigid waters while awaiting rescue. Because of the loss of life, a public enquiry was conducted and, as a result, many normally classified details were revealed by the Soviet news media.

The wrecked submarine today lies on the floor of the Norwegian Sea, about 1.5 kilometres deep, with its still-active nuclear reactor and two nuclear warheads on board. A Reuters wire release from Moscow stated, 'Russia said yesterday it had sealed a sunken nuclear submarine off Norway to prevent radioactive leaks. The *Komsomolets* ... is now embedded in mud in international waters.'

Nothing illustrates the clumsy handling of the Soviet Navy's nuclear submarines better than the events surrounding the loss of another boat, the *K-429*, better known by its NATO name as a Charlie I class. The submarine was commissioned in 1972 and joined the Soviet Pacific Fleet operating out of Vladivostock. In early 1983, *K-429* returned to base needing a major overhaul after a long patrol. The crew went on leave and handed the sub over to the dockyard, with a skeleton watch crew to keep an eye on the nuclear weapons aboard. The workers began a leisurely overhaul.

In June, however, Captain First Rank Nikolay Suvorov was ordered by the Commander of the Pacific Fleet, Rear Admiral

Oleg Yerofeyev, to prepare *K-429* for an exercise. This came as a surprise: for one thing the exercise had originally been scheduled to take place later in 1983, and the boat was spread around the dockside under repair with its crew still on leave. He was told in no uncertain terms to go ahead; the order remained unchanged. hanging over him if he refused.

Many of the crew proved unreachable, so replacements had to be found, including submariners from other vessels docked nearby. Of *K-429*'s improvised 120-strong crew, around 40 had never set foot on it before or been trained on it. The submarine was ordered to head for the torpedo firing range on 23 June 1983. Suvorov demurred, as it was protocol to perform a safety test dive first, which duly took place; but later that day, despite her dockyard condition and untrained scratch crew, *K-429* arrived at her testing area off Kamchatka. The problems began when Suvorov commanded that the sub descend to periscope depth – before the crew had taken their proper dive stations. To make matters worse, the all-important instrumentation valves remained unaligned properly, their indicators reading empty even as the main ballast tanks filled up. Assuming that the tanks weren't filling properly, Suvorov asked that the auxiliary ballast tanks be filled too. Inevitably, the excessive ballast saw the heavy boat begin to sink very quickly.

Unfortunately, workers in the dockyard had left the ventilation system wide open, to allow fumes from their welding devices to escape, and it had not been closed; and automatic locks that should have cut in to close off the ventilation valves remained switched off. With the system open at its maximum, water poured in and swiftly flooded the craft's forward compartments. Suvorov sought to blow the ballast, but the crewman responsible, one of the late, untrained recruits, mistakenly closed off the flood valves at the base of the tanks and opened those at the top, the vents on the tanks; much of the boat's high-pressure air simply whistled out of the top vents, leaving the water levels unchanged. Before the ventilation system could be shut down, the forward compartments flooded, drowning the fourteen men trapped there. The boat went down, eventually coming to rest at a depth of 40 metres.

In a catalogue of errors, *K-429*'s escape capsules and emergency buoys had been made fast – i.e. welded on to the ship's hull – for the purpose of their overhaul, so there was no safe means of exiting the sub. Suvorov and his crew decided to wait for rescue. But by the next morning, when it was obvious that nothing was happening, Suvorov called for volunteers. Two of the sub's original crew put on specialist gear and managed to swim to the surface, but found no ships there. And after they reached shore, swimming through icy water, and despite their protestations, they were promptly arrested by military police, who took them for imperialist spies.

Their desperate report eventually reached Admiral Yerofeyev at around midday on 25 June. By the early evening, the sunken submarine had been located and the evacuation began at 10:36 pm, though two of the crew were to perish during its course. The final seaman reached the surface before midnight. Captain Suvorov was one of the last to leave.

K-429 was raised on 6 August. Investigations later determined that the nuclear reactor had only partially shut down automatically, its control rods jamming in their tubes before reaching their full stroke; as a result, since the tragedy unfolded the reactor had remained operating at minimum power. However, according to the official Soviet Navy report, no radioactive leakage took place.

Captain Suvorov was scapegoated and sentenced to ten years in prison for negligence. He told an interviewer, 'I am not fully innocent. But a fair analysis should have been made to avoid such accidents in the future. I told the judges in my concluding statement: "If you do not tell the truth, others do not learn from bad experiences – more accidents will happen, more people will die."' By contrast, Admiral Yerofeyev was made the Northern Fleet's new Commander-in-Chief.

What this litany of failure and nuclear disasters at sea proves is that nuclear reactors in submarines are inherently dangerous and need very careful supervision by highly trained and experienced operators at all times. Sadly, the Soviet Navy, certainly in the early years of nuclear-powered craft, proved itself inadequate to the task. Personnel were sloppy, badly trained and prone – despite occasional heroic feats of suicidal bravery – to get things wrong in an emergency. This should give pause for thought to anyone assessing whether the Russian Navy can really be trusted with nuclear devices at sea.

After all, the uncomfortable truth is that nuclear reactors afloat are only half the problem. Even more dangerous is the cargo they transport – nuclear weapons.

CHAPTER 10

THE *KURSK* CATASTROPHE

'It's dark here to write, but I'll try by touch. It seems like there are no chances, 10%–20%. Let's hope that at least someone will read this. Regards to everybody. No need to despair.'
– LIEUTENANT-CAPTAIN DMITRI KOLESNIKOV, SUBMARINE FLEET OF THE RUSSIAN FEDERATION

On 12 August 2000, the nuclear-powered Oscar-class submarine, Project 949A, *Kursk*, sank to the bottom of the Barents Sea, with the loss of all 188 crew. The pride of the Russian Navy, it had been taking part in the country's first large-scale naval exercise for around a decade. The disaster generated a storm of publicity.

Kursk was one of the first naval vessels completed after the collapse of the Soviet Union. In many ways it was the flagship of the 'new' Russian Navy, a fleet with a long and proud reputation that it could trace back to 1696, when Peter the Great established the country's first-ever maritime force.

Ironically, the loss of the *Kursk* had nothing to do with a nuclear accident. The submarine's design actually represented the highest achievement of Soviet nuclear submarine technology. To keep up with American nuclear-powered carriers, the Soviet Oscar subs were each powered by two OK-650 nuclear reactors that together provided 97,990 shipboard horsepower. Such power gave them a top speed of 33 knots underwater – as fast as any surface ship. By any standards, these were top-of-the-range pressurised water reactors.

The cause of the accident were the torpedoes or, to be precise, the hydrogen peroxide (HP) torpedo. Hydrogen peroxide is a chemical compound (H_2O_2) created from a combination of hydrogen and water, and has had a chequered career. The Germans experimented with peroxide power towards the end of World War II. The Kriegsmarine – wartime Germany's navy – built several experimental submarines with an 'air-independent propulsion' system, using HP fuel that theoretically enabled extended underwater endurance. The Germans soon came to regard the subs as unsafe, however, as the HP fuel was dangerously volatile. After the war, Britain's Royal Navy thought the possibilities worth investigating. But when HMS *Sidon* blew up as a result of a HP torpedo malfunction and sank in 1955, killing thirteen men, peroxide drive was quietly abandoned.

Not so for the Russians. The USSR persisted with HP and, by the end of the Cold War in 1991, was fielding high-speed HP torpedoes. On 12 August 2000, *Kursk* was taking part in the country's first major naval exercise since the demise of the Soviet Union almost ten years before. But little work had been undertaken to maintain anything but the most essential front-line equipment. The bulk of the Soviet fleet was rusting at anchor, owing to a lack of funds for fuel. Consequently, many of her young, inexperienced crew had spent too little time at sea. Northern Fleet sailors of the 'Red Banner Fleet' had gone unpaid since the mid-1990s. Search-and-rescue equipment had not been maintained and the Russian submarine service was in no state for exercise, let alone war. In the five years since her launch, *Kursk* had completed only one mission, a six-month deployment to the Mediterranean in the summer of 1999 during the Balkan wars. Despite this, she was put to sea.

Kursk's state-of-the-art construction saw her lauded as almost unsinkable, rumoured to be able to cope even with a torpedo strike, while her crew had been celebrated as the finest of all the submarines in the Northern Fleet. *Kursk* was one of a select number of submarines trusted permanently to bear a combat load of nuclear missiles and torpedoes – even during a naval exercise.

Early on 12 August, *Kursk*'s captain requested permission from Northern Fleet HQ to conduct a torpedo training launch. On receiving the go-ahead, a practice Type 65 'Kit' torpedo, *sans* warhead, was loaded. The weapon was nicknamed '*Tolstushka*' ('Fat Girl') for an obvious reason: it was huge, weighing in at around five tonnes and more than ten metres long.

Something went wrong. The dummy torpedo was ten years old. Several sources later said that one of the practice torpedoes had been dropped during transport, possibly causing a crack in the casing, but that the weapon was loaded aboard the submarine all the same. What seems to have happened is that leakage of volatile HP in the torpedo tube was followed by an explosion, owing to a build-up of gas. The initial explosion was recorded by NATO submarines shadowing the Russian ships, as well as a Norwegian shore station.

Worse was to follow. Two minutes later, a huge blast, equivalent to three tonnes of TNT – recorded by seismographs as far away as Alaska – tore the *Kursk* apart. The first explosion had set off a sympathetic detonation of the submarine's full warload of torpedoes, blowing a hole in the bow. The doomed vessel plunged to the bottom in around 110 metres of water.

Another Russian submarine detected the explosion but assumed that it was part of the exercise. The Russian Task Force commander reported the explosion to headquarters, but the report was ignored. Command and control of deadly nuclear assets had broken down. By 1:30 pm there was still no contact from the *Kursk* and a helicopter was sent to look for the submarine.

The Northern Fleet duty officer notified the search and rescue forces to go to standby. By 5:00 pm, a maritime reconnaissance aircraft had been dispatched but reported nothing. An hour later, when *Kursk* failed to make her routine communications check, Northern Fleet HQ became concerned and tried to contact the ship. Something akin to panic began to set in. At 10:30 pm on Saturday, Fleet HQ declared an emergency and the exercise was stopped. A Russian nuclear submarine had sunk. Why? How?

As far as the Russian Navy was concerned, they had lost control of two top-secret nuclear reactors, plus some nuclear torpedoes and missiles. True to form, the Russians initially downplayed the incident. It seems that secretive old Soviet habits died hard. Late on Saturday night, fourteen hours after the ship sank, Northern Fleet Commander Admiral Popov ordered the first proper search for the missing submarine. At about midnight, he informed the Kremlin. There was no reaction.

Before the Kremlin had been informed of the submarine's sinking, NATO intelligence, who had been monitoring the Russian exercise closely, worked out what had happened. The British, old hands at submarine rescue and with recovery assets close by, offered to help, along with France, Germany, Israel, Italy and Norway. The United States offered the use of one of its two deep-submergence rescue vehicles. The Kremlin refused all foreign assistance, telling reporters that there was 'no big problem and a rescue was imminent'.

At 6:50 am the next day, the rescue vessel *Mikhail Rudnitsky* located *Kursk*. A Russian salvage tug obtained the first images of the wrecked submarine and unsuccessfully tried to attach to the aft escape trunk over *Kursk*'s ninth compartment. But with the weather blowing up a gale, all rescue efforts were suspended.

On Monday, 14 August, Russia made its first public statement about the incident, downplaying *Kursk*'s plight as one of trifling technical problems. It asserted that the sub was now on the seabed but that the crew had been contacted, were in good health and were being supplied with air and power. It was all a lie.

Poor weather conditions hampered operations over the following two days, preventing the rescue team from attaching a diving bell to *Kursk* or attach a remotely operated vehicle to its rescue hatch. On Tuesday, three days after the sinking, the rescue ship managed to lower a diving bell, but were unable to connect it to the sub. The remotely operated vehicle (ROV) *AS-34* was damaged while attempts were being made to fasten it to *Kursk*'s rescue hatch, and had to be repaired, further delaying proceedings. That same day, the Russian crane ship *PK-7500* brought a Deep Submersible Rescue Vessel (DSRV), but again the poor conditions thwarted its launch. The Wednesday saw failed attempts to attach the *AS-34* or the DSRV to one of the submarine's escape hatches.

By now *Kursk* was headline news around the world. Even state-owned Russian newspapers criticised the navy's refusal to accept international assistance. Various explanations for the accident were put forward in public by senior Russian officers. Four days after *Kursk* sank, the Russian Navy Commander-in-Chief, Fleet Admiral Kuroyedov, said that the accident had been caused by an underwater collision.

By Thursday, the navy staff was still maintaining their understanding that no explosion had occurred on the *Kursk*, and that that the sub was intact on the seabed. The Kremlin were dependent on these reports for their information about what had happened. Desperate attempts by the DSRV to reach the escape trunk in the aft area of the submarine foundered. Moscow even issued a false story to the media that rescuers had heard tapping from within the boat's hull. Finally, on 17

August 2000, President Putin agreed to foreign offers of assistance and, on 19 August, the Norwegian ship *Normand Pioneer* arrived with the British rescue submarine *LR5* on board. It was now seven days after the disaster. On Sunday, 20 August, the Norwegians lowered an ROV to the submarine. They reported that a major explosion had taken place and that the bow of the submarine had been torn wide open.

On Monday, 21 August, after the Norwegian divers confirmed that no one was alive in the ninth compartment, the Russian Northern Fleet finally announced to the public that the *Kursk* was flooded and all its crew members lost.

The blame game started early. The official line from the Kremlin was that the accident had been caused by a serious collision with a NATO submarine. The Russian government convened a commission of enquiry, and chaired by Vice Premier Ilya Klebanov. It was rigged, though, consisted of naval officers and officials with a vested interest in hushing up the incident and blaming 'foreign powers'. There were no independent investigators. It was a classic Communist-era cover-up.

Senior commanders of the Russian Navy continued to parrot this false account for another two years. In late August 2000, the official government commission reported that a 'strong dynamic external impact' – such as a foreign submarine or even a mine left over from World War II – had probably caused the sub to explode. Admiral Popov gave an interview to the Spanish investigative newspaper *El Mundo* on 25 October 2000 in which he reiterated that *Kursk* had most likely struck a NATO sub that had been monitoring the exercise, which he stated had been observed by two American submarines, USS *Memphis* and USS *Toledo*, as well as the Royal Navy submarine HMS *Splendid*. In due course, the Russian Navy released satellite images of the *Memphis* docked in Norway, supposedly to undergo repairs for the damages sustained during the collision. The usual crop of conspiracy theories circulated: Chechen sabotage, human error and even that the *Memphis* had fired a torpedo at *Kursk*.

International geophysicists were adamant that *Kursk* had been rocked by two explosions from torpedo detonations; a British seismic monitoring station also identified two distinct bangs consistent with exploding torpedoes. Despite this, Russia persisted with its attempt at a cover-up, going so far as to withhold from the families of the crew the full list of missing sailors. After considerable effort, a *Pravda* reporter was eventually able to get the 'state secret' list.

Things became even worse for the Russian version of events when TV pictures of distraught wives and parents of the crew went international. On Tuesday, 22 August, ten days after the sinking, President Putin met angry family members of the *Kursk*'s crew. Furious women barracked him, demanding to know why they were receiving such conflicting information and asking who was going to be punished for the deaths of their family members.

The mother of *Kursk* submariner Lieutenant Sergei Tylik, became particularly emotional, berating Putin and Deputy Prime Minister Klebanov and declaring, 'You better shoot yourselves now! We won't let you live, you bastards!' When she would not quieten down, a KGB/FSB nurse injected Madame Tylik through her clothing with a powerful drug. She collapsed and was carried out, to the astonishment of the watching journalists and a German TV crew who had filmed the entire incident. A Russian journalist, Andrey Kolesnikov, who had witnessed Putin trying to fob off the families, said he had never felt such an intense atmosphere. 'I honestly thought they would tear him apart,' he said. 'There was such a heavy atmosphere there, such a clot of hatred, and despair, and pain . . . I never felt anything like it anywhere in my entire life . . . All the questions were aimed at this single man . . . ' Taken aback by the onslaught of accusations against Putin and his government, Defence Sergeyev and senior commanders of the navy and the Northern Fleet tendered their resignations. A furious Putin turned them down.

The final miserable truth was confirmed two years later. On 26 July 2002, Russia's Prosecutor General, Vladimir Ustinov, announced that the cheap HP fuel in the dummy torpedo within the fourth torpedo launcher had ignited the first explosion that was to sink *Kursk*. The following month, the government published a summary of an official classified report in Rosslyskaya Gazeta confirming that *Kursk* had been sunk by a torpedo explosion caused when highly concentrated HP leaked from cracks in the torpedo's casing. Ustinov acknowledged there had been 'stunning breaches of discipline, shoddy, obsolete and poorly maintained equipment' along with 'negligence, incompetence and mismanagement'. The report also admitted that the rescue operation 'had been unjustifiably delayed'. Retired Russian Navy Captain Vladimir Mityayev lost a son on *Kursk* and observed, 'To me, this is a clear case of negligence.'

In the end, in true Soviet style, no one was blamed for the disaster and no one was held responsible. It was, however, the signal for all change in the Northern Fleet. Putin appointed a new submarine commander, Vice Admiral Oleg Burtsev, and replaced twelve high-ranking Northern Fleet officers.

Britain's *Guardian* newspaper summed the whole sorry saga up in a savage judgement in 2002:

> The hopelessly flawed rescue attempt, hampered by badly designed and decrepit equipment, illustrated the fatal decline of Russia's military power. The navy's callous approach to the families of the missing men was reminiscent of an earlier Soviet insensitivity to individual misery. The lies and incompetent cover-up attempts launched by both the navy and the government were resurrected from

a pre-Glasnost era. The wildly contradictory conspiracy theories about what caused the catastrophe said more about a naval high command in turmoil, fumbling for a scapegoat, than about the accident itself.

Eventually the wreck of the Kursk was salvaged by two specialist Dutch marine salvage companies in what became the largest operation of its kind, and one of the most dangerous because of the risk of radiation from the two nuclear reactors. Only seven of the submarine's twenty-four torpedoes were accounted for. The location of the remainder, along with the presence of seven unexploded torpedo warheads, the twenty-two SS-N-19 *Granit* cruise missiles aboard, plus a missile ejection charge in each silo, remains unknown. Looking at the wreck, it is quite clear that there had been a catastrophic explosion in the forward torpedo compartment; the bow of the submarine had been severely damaged. What was left was cut off during the salvage operation.

President Putin later signed a decree awarding the Order of Courage to the entire crew, and the title Hero of the Russian Federation to the submarine's captain, Gennady Lyachin. The sail of *Kursk* was rescued from a scrapyard and now stands as a memorial at the Church of the Saviour on Waters overlooking Murmansk. It is dedicated to the men who died aboard the sub, 'To the submariners, who died in peacetime'.

What to make of the *Kursk* disaster? It was not directly a nuclear accident but it drives home a powerful lesson – that there are three absolutely basic principles regarding the existence and use of nuclear weapons:

- Safety
- Centralised command and control
- Protection

This is reinforced by the International Atomic Energy Authority's wider definition of nuclear security and protection as: 'The prevention and detection of, and response to, theft, sabotage, unauthorised access, illegal transfer or other malicious acts involving nuclear material, other radioactive substances or their associated facilities.'

Through incompetence, cost-cutting, bad training, sloppy procedures and a complete breakdown of command, control and communications with nuclear assets, the Russian Navy lost control of two highly dangerous nuclear reactors in the Barents Sea, along with *Kursk* and its full crew. Then, unforgivably, it tried to cover up its loss and its mistakes.

But you dare not take risks with nuclear. In 2000, the Russian Navy gave a masterclass in just how *not* to handle nuclear submarines.

CHAPTER 11

MANNA FROM HEAVEN? LOST BOMBS

'Manna exists and is common in the Sinai desert. A plant, *Haloxylon salicornicum*, is found all over the Middle East. The white drops are really the digestive by-product of insects, known as honeydew.'
– **PROFESSOR AVINOAM DANIN, HEBREW UNIVERSITY OF JERUSALEM**

'Manna' is first mentioned in the Bible, when God promises Moses to 'rain bread from heaven' to feed the Israelites. Next day, the desert rocks are coated with sweet, sticky food. Unfortunately, in the nuclear age, not all the stuff raining down from heaven is quite so welcome...

In 1978, the Canadian authorities received an unwelcome gift from above. On 24 January, *Kosmos 954* – a Soviet radar ocean reconnaissance satellite with an onboard nuclear reactor – broke up on re-entry over the distant Northwest Territories. The nuclear fuel was spread over a wide area and some radioactive pieces were recovered.

The *Kosmos 954* incident brought into sharp public focus the reality of a nuclear accident from space. Ocean reconnaissance satellites are invaluable tools for keeping an eye on shipping. You simply cannot hide a ship alone on an empty ocean. The early Soviet low-orbit radar satellites were unique. They used miniature nuclear reactors as a power supply because their orbits took them through the upper atmosphere, whereas the drag from solar power panels – more usually associated with satellites – would pull the satellite back down to Earth. To get rid of any radioactivity once a satellite's working life was ended, Soviet engineers designed the satellites to eject their reactor cores into a higher, eternal 'graveyard orbit'. It seemed a good idea at the time. But, as in all human endeavours, things go wrong.

The North American Aerospace Defense Command (NORAD) is a combined US and Canadian organisation that provides aerospace early warning for Northern America by means of powerful tracking radars and satellites. From its headquarters in the Cheyenne Mountain Complex, its radars track anything coming into North American airspace, at whatever altitude, particularly over the North Pole. Rumour has it that it can spot loose space junk, and even migrating flocks of geese, let alone incoming ballistic missiles. On Christmas Eve it even supposedly tracks Santa, ever since a little girl dialled its unlisted phone number back in 1955, thinking she was calling Father Christmas.

Satellites normally don't act erratically, so by late 1977, NORAD's command staff were beginning to take a close interest in one particular Soviet satellite, *Kosmos 954*, which was behaving unusually. In the 1970s, a satellite making lots of manoeuvres in a short time, or inexplicable sudden orbit changes, was worrying.

A satellite that suddenly loses fifty kilometres of altitude looked odd. In 1973, a previous Soviet low-orbit satellite had failed, dumping its nuclear reactor into the Pacific Ocean north of Japan. In December 1977, the US National Security Council was alerted that the *Kosmos 954*'s orbit was over the USA. The Kremlin's specialists confirmed that they had lost control of their satellite. The system intended to propel the spent reactor into a disposal orbit had failed – its enriched uranium-235 core might fall to Earth on US soil.

On 24 January 1978, *Kosmos 954* began its final descent. The errant satellite re-entered Earth's atmosphere and NORAD reported that it was inbound on a trajectory for Canada's remote Northwest Territories. At first, the USSR claimed that the satellite had been completely burned up and destroyed on re-entry. This was yet another Soviet Cold War lie, the Kremlin's knee-jerk denial of anything potentially embarrassing. It subsequently emerged that debris from the satellite's wreckage had crashed to Earth and scattered radioactive fragments across 640 kilometres of the region.

A massive winter search operation ensued in the frozen north. Joint Canadian-American teams swept the area on foot and by air with hand-held Geiger counters. What they found was alarming. Just twelve large pieces of the satellite were discovered in the snow, of which ten were radioactive. That only accounted for 1 per cent of the nuclear fuel, but one fragment had a radiation level of 500 rads per hour, 'sufficient,' as stated in *Space Law in the United Nations* (1985) by Marietta Benkö, 'to kill a person ... remaining in contact with the piece for a few hours'.

Many of the pieces landed on the top of frozen lakes, later to sink to the bottom in the spring melt. Moscow declined Canada's offer to take back the debris and instead paid out $3 million under the terms of the Space Liability Convention, which obliges a state that launches an object to pay for any damage caused.

Five years later, in 1983, a similar Soviet satellite, *Kosmos 1402*, suffered an alarmingly similar malfunction. The main pieces plunged into the South Atlantic. After that, Soviet low-orbit satellites were re-equipped with a system that allowed their nuclear-reactor section to be ejected into a high-disposal orbit,

960 kilometres out in space, where they remain as deadly space junk until their radioactivity decays in 500 years.

The truth is that hundreds of hazardous objects have dropped from the sky ever since nuclear hardware went airborne. There have been too many accidents involving aircraft dropping nuclear weapons or crashing with bombs on board. One such event happened on 10 March 1956, when a USAF B-47 descended into clouds at 4,270 metres (14,000 feet) over the Mediterranean . . . and vanished. Unfortunately, it was carrying two nuclear weapon cores. No wreckage has ever been found.

There are many more examples from which to choose – far too many. Recent history is littered with nuclear accidents. The following is a selection of some of the most notorious reports of nuclear bombs arriving inadvertently from the sky since that B-47 incident. They make for alarming reading.

In early 1958, at Greenham Common air base in England, a B-47E of the USAF 310th Bomb Wing developed problems on take-off and was forced to jettison two external fuel tanks. One hit a hangar at the end of the runway while the other struck the ground twenty metres behind another parked B-47E preparing for take-off. The stationary plane, which had just been fuelled, was loaded with a 1-megaton nuclear bomb. Suddenly it was engulfed by flames. The resulting fire took sixteen hours and more than three million litres of water to extinguish; controlling the fire becoming a major problem because the aircraft was built using magnesium alloy, which burns at over 3,000°C. The US and UK authorities kept the potential nuclear accident very quiet. Even by the mid-1980s, the British government's story remained that a parked aircraft had been struck by a taxiing aircraft and that there was no fire, despite the fact that two men died and eight were injured in the event.

That reaction was (and probably still is) the default position on the military's numerous nuclear accidents, using the convenient excuse of secrecy and national security. However, sometimes accidents are just too public to be brushed away by official denials, obfuscation or spin. In 1961, one incident left an unexploded nuclear bomb in rural America in what came to be known as the Goldsboro B-52 crash. Despite the tragedy of the night, it has elements of farce too.

In the early hours of 24 January 1961, bright orange flames lit up the night sky over North Carolina. Inhabitants of the small farming community of Faro saw the final moments of a USAF B-52G jet, with two hydrogen bombs on board, as it disintegrated in mid-air. The plane was one of a dozen Strategic Air Command bombers in the air that night, ready for an instant retaliatory strike should the Soviet Union attack the USA. Suddenly, without warning, the B-52 lost seventeen tonnes of fuel in just two minutes. The baffled pilots decided on an emergency landing at Seymour Johnson Air Force Base near Raleigh. Third pilot, First Lieutenant Adam Mattocks, was ordered to the controls and told to level the jet off at 3,050 metres (10,000 feet) and hold her straight and level while the puzzled senior pilots and the flight engineer tried to work out what had happened, checking the complex of fuel gauges.

Then the right wing broke off. The massive bomber promptly dropped into a rapid right-hand spin and now, trailing fire, plummeted towards the ground. As the B-52 nosedived and spun to the ground, Mattocks tried to bail out, but the pressure of the g-force from the spin pinned him to the floor. Terrified, the twenty-seven-year-old Carolinian prayed, 'Lord, if I go, take me home to heaven.' Above him the co-pilot managed to force open his hatch and jump out of the plane. By a superhuman effort, Mattocks crawled across the whirling floor of the flight deck and flung himself out into the darkness. As he parachuted into the night, he saw the flaming wreckage of the B-52 'spinning like a Catherine wheel' to explode far below.

One astonished local witness said that the explosion in the sky 'lit the place up like daylight'. A woman dropped to her knees and prayed, certain that the second coming had arrived. Another man reported that he thought that the burning bomber had crashed on to his parents' house. Billy Reeves, who was eighteen at the time, had just gone to bed when he heard what he later described as 'a weird sound'. Suddenly his room was lit up. He ran to the window to see a flaming aircraft coming down to crash and explode near Faro.

Earl Lancaster, the assistant fire chief for the Faro Volunteer Fire Department, rushed to the scene in his fire truck. Whole fields were on fire. Then, within an hour, helicopters swarmed into the area, disgorging armed US Marines and air-force officials who urged everyone to evacuate.

Although five of the crew members parachuted to safety that night, three men died. The body of forty-one-year-old Major Eugene Shelton, a radio navigator, was found three kilometres from the crash site, hanging upside down from a tree by his red-and-white parachute. The bodies of the electronics weapons officer, Major Eugene Richards, and gunner Sergeant Francis Barnish were found in the nose of the plane. Lieutenant Mattocks landed beside a nearby farmhouse, shaken but unscathed. He pulled off his mask and told the family there what had happened. Like good citizens, they drove him to the nearby air base. But once he arrived at the gate, Mattocks realised that he had no identification – the g-forces had ripped his pockets off. He tried to explain what had happened but the suspicious MP guards refused to believe him. To them, he was just another guy dressed in an old USAF military uniform trying to sneak into an off-limits top-secret nuclear air base. They arrested the stunned and protesting pilot. An hour or so later, the bizarre situation grew even more surreal. The aircraft's captain, Major Walter Tulloch, startled everyone by suddenly walking out of the swampland, dazed but, apart from a ripped flying suit, surprisingly calm. He made his way to the air base and – battered, dishevelled and angry – presented himself at the gate. He, too, had no identification. The MPs promptly arrested him as well.

Eventually, the situation was resolved after air-force officers arrived to confront two very angry pilots and a nuclear disaster on their doorstep. Thanks to the painstaking research of Harrison Williams of Stanford University, we now know more of the technical details of what really happened. Two of the latest USAF Mark 39 nuclear bombs, each (200 to 300) times more powerful than the bombs dropped on Hiroshima and Nagasaki, were lying somewhere on US soil, unsecured and liable to explode at any moment. The force of the B-52's mid-air break-up almost certainly initiated the fusing sequence for both bombs. Panic set in. American nuclear UXBs? Dropped on American soil? For the Pentagon, this was the stuff of PR nightmares.

When responders arrived at the crash scene, they discovered that one of the nuclear weapons had landed with its deployed parachute tangled in the branches of a tree and its nose stuck in the ground. It was deactivated, loaded on to a truck and spirited away to be examined.

Unfortunately, the second bomb had anything but a soft landing. Its parachute had not deployed, the live H-bomb breaking up as it buried itself deep in a field. Its plutonium and uranium core was located somewhere six metres below the surface. To their horror, investigators saw among the wreckage that the bomb's arm/safe switch was on 'armed', as a result of the impact. In a secret document dating from 1969 (but only declassified in the twenty-first century), Parker F Jones, supervisor of the nuclear weapons safety department at Sandia National Laboratories, revealed that of the four safety mechanisms designed to intervene in accidental cases such as this, just one had remained locked – the only thing stopping a nuclear detonation.

With unusual bureaucratic levity, Jones wryly re-entitled his official report *Goldsboro Revisited or How I Learned to Mistrust the H-Bomb*, a nod to the full title of Stanley Kubrick's 1964 nuclear satire *Dr Strangelove*. Jones's report concluded that, 'When the bomb hit the ground, a firing signal was sent to the nuclear core of the device, and it was only that final, highly vulnerable switch that averted calamity.' He added, 'One simple, dynamo-technology, low-voltage switch stood between the United States and a major catastrophe.' The safety supervisor also pointed out that if the mid-air breakup of the B-52 had caused an electrical short circuit in the doomed switch, it could have set off a nuclear explosion bigger than Hiroshima and Nagasaki combined. Jones concluded that, 'The Mk 39 Mod 2 bomb did not possess adequate safety for the airborne alert role in the B-52,' and that the devices designed to prevent an accidental detonation were, 'not complex enough'.

A worried air-force team dug in the marshy field for missing parts of the bomb. After six metres, they hit the water table. The recovery team had sixteen pumps, capable of sucking out several thousand litres of water an hour, but the flow was relentless. Defeated, they hastily filled in the hole, leaving parts of the bomb deep in the Carolina farmer's field to this day.

While the fact that the crippled B-52 was carrying two nuclear weapons was widely reported at the time, the military were rather more circumspect about just how close the accident had come to causing a nuclear catastrophe. The official account from the Pentagon states that there 'was no hazard in the area'. However, safe or not, the US government still quietly collects samples from wells near the crash and the Pentagon purchased the land from the farmer and his heirs. It's still down there. Somewhere.

The Goldsboro near-disaster resulted in even more stringent safeguards being placed on America's nuclear arsenal. Experts estimate that had one of the bombs involved in the 1961 crash detonated, the blast would have instantly killed everything within an thirteen-kilometre radius. Lethal radiation fallout would have been borne on the prevailing winds up the Atlantic seaboard, perhaps as far as New York.

While it's a particularly dramatic example of a broken arrow, the Goldsboro accident serves as just one of the many nuclear accidents involving aeroplanes. That aircraft occasionally crash is a fact of life. We don't have to look very far to uncover other examples. On 14 March 1961, just two months after the explosion in Goldsboro, a USAF B-52 bomber took off from Mather Air Force Base in California. Twenty minutes later, the pilot called in to complain of scorching hot air filling the cockpit. Then his window blew out. Faced with the loss of cabin pressure, he decided to bring the aircraft down to 10,000 feet (3,050 metres). Inexplicably, it ran out of fuel and crashed near Yuba City in California with its two nuclear bombs still on board. Fortunately, there was no explosion. The Pentagon and Atomic Energy Commission rushed to clean up the site.

Fifteen years later, the US government released pictures of the Yuba crash scene. The silent movie shows debris, with troops taking radiation readings of the wreckage. Airmen wander through the B-52's scattered remains, picking over debris without even using gloves, despite the danger of radioactive contamination. The fact that no one is wearing protective gear shows how casually the military treated these accidents during the early decades of the Cold War.

Three years later, Frostburg, Maryland, was the reluctant recipient of yet another unwelcome dose of nuclear manna from the skies. On the night of 13 January 1964, a B-52 Stratofortress was on an airborne-alert training mission. Things started to go wrong when, at more than 30,000 feet (9,150 metres), the eight-engine bomber began to shudder violently. The B-52's pilot, Major Thomas W McCormick, later explained, 'I encountered extreme turbulence . . . the aircraft became uncontrollable and I ordered the crew to bail out.' The bomber's tail stabiliser broke off and the crippled aircraft spun out of control. Only the two pilots survived in the freezing forest. One crew member failed to get out and two more died from injuries or exposure to the harsh winter weather after landing. A five-day search for the missing weapons eventually succeeded in finding and recovering the two nuclear bombs.

On 4 July 1964, the survivors, USAF pilots Peedin and McCormick, attended the dedication of an American Legion memorial to the downed fliers at Big Savage Mountain. Dipping its wings in salute, a lone B-52 bomber flew low overhead. Nowadays, hikers stop at the mossy stone cross by a gentle stream, not realising they are standing on the site of a near nuclear catastrophe.

In 1965, the US Navy made its contribution to losing nukes when a Skyhawk with one B43 nuclear bomb had a 'cold cat' – i.e. the steam catapult fails to supply enough forward momentum for a successful take-off – from the aircraft carrier. It dropped into 4,900 metres (16,200 feet) of water about 320 kilometres from Okinawa while the ship was underway from Vietnam to Yokosuka, Japan. The incident was hushed up and did not come to light until a 1989 US Department of Defense report revealed that a 1-megaton bomb had been lost. That weapon was never recovered.

The accidents kept piling up, and some of them had international consequences. On 21 January 1968, a fire broke out in the navigator's compartment of a USAF B-52 near Thule Air Base in Greenland. The aircraft was flying as part of the SAC 'Chrome Dome' operations, which kept nuclear bombers loitering airborne 24/7, ready to attack at the first sign of a Soviet missile launch against the USA. The fire spread and the bomber crashed, smashing on to sea ice in Greenland's North Star Bay; the impact caused the conventional explosives aboard to detonate and the nuclear payload of four hydrogen bombs to rupture and disperse, scattering radioactive contamination. The

United States and Denmark (Greenland was a Danish territory at the time) launched an intensive clean-up and recovery operation, complicated by Greenland's harsh weather. After the operation was completed, the secondary stage of one of the nuclear weapons remained accounted for.

In the spring, contaminated ice and debris were flown to the United States for disposal. The incident caused protests in Denmark, as Danish law forbids nuclear weapons on its territory. For years afterwards, the Danish clear-up workers, who had in the meantime begun to suffer from seriously raised cancer rates, tried to sue the USA, but without success. The whole incident led to bad blood between two NATO allies. Shortly afterwards, the USAF Strategic Air Command Chrome Dome operations were discontinued.

Perhaps the most public, if not the most dangerous, broken arrow incident was an infamous debacle in 1966, when a nuclear bomber of the USAF Strategic Air Command dropped four hydrogen bombs over eastern Spain. Three were found on land near the small fishing village of Palomares. The fourth disappeared into the Mediterranean Sea. The non-nuclear explosives in two of the weapons detonated upon impact with the ground, resulting in plutonium contamination across an area of two square kilometres.

In 1966, the USAF's Strategic Air Command was keeping at least a dozen B-52s patrolling the skies over the Atlantic and Europe around the clock, each with a payload of hydrogen bombs. On 17 January 1966, a B-52G commanded by Captain Charles Wendorf was several hours into a routine airborne alert flight and headed for a refuelling over Spain. At around 10:20 am, relief pilot Major Larry Messinger manoeuvred the B-52 underneath and behind a KC-135 Stratotanker, aiming to connect the fuel probe on the B-52 with a refuelling boom extended from the tanker. As the two aircraft drew close, Messinger recalled, 'All hell seemed to break loose.' The bomber's incoming speed was too fast and caused the aircraft to collide with the tanker's fuelling boom. The belly of the KC-135 was sliced open, and jet fuel spilled out, drenching both planes. Explosions ripped through the two aircraft, consuming the tanker and killing all four men aboard. Three men in the tail of the bomber were killed, and the four other crew members bailed out. Messinger recounted, 'I opened my parachute. Well, I shouldn't have done that. I should have free-falled and the parachute would open automatically at 14,000 feet [4,270 metres],' he said. 'But I opened mine anyway, because of the fact that I got hit in the head, I imagine.' He drifted east, thirteen kilometres out to sea, where he was picked up by fishermen. However, as the B-52 broke up, its unarmed thermonuclear payload of four 1.5-megaton bombs fell out. Three of the hydrogen bombs hit the ground, while the fourth landed in the sea. Wreckage from the B-52 and KC-135 smashed into Palomares, a small agricultural community previously best known for growing tomatoes. 'I looked up and saw this huge ball of fire, falling through the sky,' Manolo Gonzalez, one of Palomares's two thousand residents, later told Public Radio International. 'The two planes were breaking into pieces.'

Flaming wreckage slammed into the ground, some near the local schoolhouse, although no one in the village was killed. The four surviving American airmen were quickly found and taken to a nearby hospital. At that point, none of the villagers knew that somewhere in the debris littering the fields around their homes were three Mark 28 hydrogen bombs.

None of the bombs produced a nuclear blast, but the conventional explosives on two of them detonated upon impact, dispersing radioactive plutonium fragments across the countryside around Palomares. Something akin to panic erupted at USAF HQ in Spain.

'It was just chaos,' John Garman, then a military police

officer, told *The New York Times* in 2016. 'When we got there, wreckage was all over the village. A big part of the bomber had crashed down in the yard of the school.' US personnel began a frantic search for the four missing nuclear bombs the B-52 had been carrying. The weapons – each capable of producing an explosion about a hundred times greater than the bomb dropped on Hiroshima – were not armed, so there was no chance of a nuclear detonation. Radiation contamination was another matter.

Although one bomb was discovered intact, in two others the high explosives designed to initiate a nuclear blast had detonated on impact, scattering plutonium and contaminating crops and farmland round the village. Thompson and his colleagues spent days searching the contaminated fields without protective equipment or even a change of clothes. 'They told us it was safe, and we were dumb enough, I guess, to believe them,' he reflected. In early February, *Life* magazine described the previously sleepy coastline as looking like 'a World War II invasion beachhead', crawling with US servicemen.

Despite early reports in the media that a nuclear device had been lost, the US military and the Spanish regime kept quiet about what was going on at Palomares. One US briefing officer stated, 'I don't know of any missing bomb, but we have not positively identified what I think you think we are looking for.' This feeble attempt at a cover-up only attracted Soviet attention, with Radio Moscow broadcasting that the coastal countryside was heavily contaminated with 'lethal radioactivity'.

The real problem was that missing fourth bomb. The USAF had to locate it for several reasons, including security, safety, and the threat to Spanish fishermen and villagers. Above all, its loss was deeply embarrassing for the USA, so much so that its ambassador to Spain was reluctantly dragooned into a PR stunt on 8 March, going for a swim to prove the water was safe. He told reporters, 'If this is radioactivity, I love it!'

A fisherman who had seen the bomb come down out at sea then yielded a vital clue. What a Spaniard had thought was a dead man on a parachute could have been the missing weapon. The search switched to a new area. The navy brought in *Alvin*, a submersible deep-sea search vessel capable of diving to 1,830 metres and equipped with cameras and a grappling arm. Two weeks later, *Alvin* found the missing bomb and attached a line to winch it to the surface. Finally, on 7 April, the US Navy recovered the USAF's missing hydrogen bomb, bringing the eighty-one-day saga to a close. *The New York Times* reported that it was the first time the US military had put a nuclear weapon on public display.

The row in Spain over the incident caused serious international tensions. The US government paid $710,914 to settle 536 Spanish claims. The fisherman whose report had led to the discovery of the missing bomb wanted a salvage claim and sued for $5 million, but was eventually awarded a paltry $14,566. In Madrid, protesters had chanted 'Yankee assassins!' during the search, and the Spanish government asked US Strategic Air Command to stop its flights over the country. In the south, as part of 'Operation Moist Mop', the Americans removed 400 tons of plutonium-contaminated soil from 260 hectares of land, sending it to an approved storage facility in Aiken, South Carolina.

Since then, the sorry saga has dragged on. A 2006 State Department cable revealed that the Spanish government 'believe[d] the remaining contamination might be more serious than heretofore believed' and that the US had spent $12 million, or approximately $300,000 every year, to assist the Spanish in monitoring the contaminated area. In 2006, the Spanish Centre for Energy Research reported radioactive snails in the area. Subsequent analyses continue to detect levels of plutonium in the soil and part of the area still remains fenced off. In October 2015, after several years of negotiations, the US government signed a statement of intent to continue assisting Spain in finishing the fifty-year-old clean-up process in Palomares.

In recent years, a number of US servicemen who participated in the clean-up have claimed that their exposure to plutonium has resulted in lifelong health problems. One is Victor Skaar, late of the USAF. For sixty-two tough days, when food was scarce and US servicemen slept in tents, Skaar took measurements in Palomares and helped collect contaminated soil in barrels. Upon his return to the USA in late 1966, the air force told Skaar that he would undergo urine tests for the rest of his life to determine radioactivity levels. In 1982, he was diagnosed with leukopenia, a condition that reduces white blood cells. Later, he suffered from prostate cancer and skin cancer. Doctors believed radioactivity was the most likely cause. Skaar applied for disability benefits from the Department of Veteran Affairs but, inexplicably, his request was denied. According to the US Armed Forces, neither he nor the nearly 1,600 soldiers who were in Palomares were exposed to radioactive risks. Their USAF medical records have long disappeared.

Palomares remains a live issue to this day. It is a classic example of nuclear going wrong and the consequences of accidents. That long-ago crash and potential nuclear catastrophe casts a lengthening shadow and still poses important questions about the safety of nuclear weapons for us today. One simple accident like an aircraft crash, just one error in the ageing software controlling our nuclear weapons, or even a basic mechanical fault within one of those weapons, retains the potential to raze the countryside, destroy entire cities or even start a war. That may sound like hyperbole; it is simply the truth. We can try to ignore the danger, but it still is lurking out there, waiting to fall from the sky.

And, unlike manna, it's not a blessing from heaven.

CHAPTER 12

FIREWORKS CAN BE DANGEROUS

- Only grown-ups should buy and let off fireworks.
- Stand well back.
- Keep fireworks in a closed box.
- Light at arm's length, using a taper.

**– ROYAL SOCIETY FOR THE PREVENTION OF ACCIDENTS –
ADVICE FOR FIREWORKS**

On 18 September 1980, an airman conducting maintenance on a USAF Titan-II missile at Little Rock Air Force Base in Damascus, Arkansas, dropped a socket wrench, which fell about twenty-four metres, piercing the skin on the rocket's first-stage fuel tank, causing it to leak. The area was evacuated. Overnight, at about 3:00 am, the hypergolic fuel (in plain English, rocket fuel that will spontaneously ignite when mixed with another substance) exploded. The missile blew up in its launch tube. Its W53 nuclear warhead popped out like a champagne cork and landed about thirty metres from the launch complex's entry gate, much to everyone's surprise. Fortunately, its safety features operated correctly and prevented any explosion, conventional or nuclear, although Senior Airman David

Livingston was killed and the launch complex was destroyed. Stunned USAF personnel tried to work out what to do next. This was not in the 'USAF Nuclear How to Do It Manual'. Even on the ground, nuclear missiles are clearly extremely dangerous, but this disaster is merely one of a long list of things that have gone wrong with nuclear weapons on the ground. Far worse was the Soviets' failure of a new ICBM on 24 October 1960.

The 'Nedelin Incident' at the Tyuratam test range in southern Kazakhstan, now part of the Baikonur Cosmodrome facility, remains one of the worst nuclear-related accidents on record. A prototype of the new Soviet ICBM R-16 missile was being made ready for a test flight when an explosion occurred following accidental ignition of the second-stage engine. An unknown number of personnel working on or near the rocket were killed. News of the accident was suppressed by the Soviets for nearly thirty years.

The event is named after the head of the R-16 ICBM development programme, Chief Marshal of Artillery Mitrofan Ivanovich Nedelin, who died in the incident. In 1960, Soviet leader Nikita Khrushchev was pushing hard for a public demonstration of the Soviet Union's Cold War superiority over the West. The new R-16 ICBM rocket promised to be a powerful symbol of the USSR's might and engineer Mikhail Yangel, along with his design team, aimed to give Krushchev a major publicity coup on 7 November for the celebration of the Bolshevik Revolution: a successful test launch of the first R-16. It could deliver a single 1,900-kilogram nuclear-tipped re-entry vehicle to a range of 11,112 kilometres (6,000 nautical miles).

In September 1960, the first rocket was delivered to the test range in Tyuratam. The design team was uneasy about its hasty preparation. On 26 September, however, the project experienced so-called 'technical problems' and an ad hoc state commission was formed to oversee the tests and to spread responsibility among Soviet bureaucrats.

By 20 October, technical tests were declared complete and the missile was moved from the processing hangar to the launch pad. Despite advice from experts that the rocket was still plagued by problems and not ready for launch, the two-stage, liquid-propellant intercontinental ballistic missile was prepared for lift-off. All that test flight lacked, mercifully, was its nuclear payload.

Pre-launch checks of the eight-storey-tall missile took two days and by 23 October it was ready to be fuelled. The dimethylhydrazine fuel in combination with red fuming nitric acid oxidiser fuel was nicknamed 'devil's venom' because the liquid-oxygen fuel mix is extremely corrosive and highly reactive, producing poisonous gas when burned. In effect, the R-16 was a giant, liquid-fuelled firecracker.

All unessential personnel were supposed to leave the area prior to the launch, but Marshal Nedelin and the missile's chief designer, Mikhail Yangel, opted to stay at the pad. Around 150 others remained clustered round the base of the rocket.

Ominously, the launch crew located a fuel leak on the rocket during, or not long after, fuelling; it was producing around 145 drops per minute. No system existed for draining off the propellant, though; and even if one had been in place, the missile couldn't be reused for a second launch once these corrosive fluids had been removed. Instead, the technical management dismissed the leak as 'acceptable', ordered members of a chemical unit to keep it under control and authorised the launch procedure to continue. Now loaded with its volatile liquids, the R-16 missile and its launch team had effectively reached the point of launch or abort.

Technicians sent a command to initiate the membranes on the oxidiser line of the rocket's second stage. Owing to defects in their design and manufacture, however, they malfunctioned. With the membranes closed, and the corrosive liquid fuel already eating away at the pipes and tubing, the rocket could only stay at the launch pad for a couple of days; the window for the launch was getting narrower. The situation deteriorated further: pyrotechnic devices controlling the valves of one of three engines in the rocket's first stage ignited spontaneously. Then the electrical current distributor that supplied the power to the rocket failed.

Major General Gerchik, head of Baikonur Cosmodrome, would subsequently claim that at this point his personnel proposed that the fuel be drained and the rocket removed. It's possible, as some accounts suggest, that Marshal Nedelin angrily insisted that they wouldn't have the opportunity for such niceties in a real-life nuclear confrontation. The Kremlin wanted results.

By the early evening on 23 October, pre-launch work came to a halt, so that the malfunctioning valves and power distributor could be replaced. The attempt was re-scheduled for the following day.

Thanks to a NASA briefing on the disaster in 2005, we know what really went wrong that day. On the morning of 24 October, preparations for the launch restarted, with Boris Konoplev, the control system's lead designer, overseeing checks on the launch pad itself.

The clock ticked down. Representatives of the state commission assembled at ground control, not far from the pad. Marshal Nedelin was apparently now fielding calls directly from the Kremlin, demanding to know precisely when the launch would take place. When another half-hour delay was announced, Nedelin stormed down to the launch pad and was seated some eighteen metres from the rocket. The state commission members joined him there.

Politics and the looming launch deadline were piling pressure on the ground personnel. An engineer who had a basic plan of the wiring was banned from the launch pad, allegedly for security reasons. Given the harried atmosphere, it's inevitable that serious mistakes were made. A switch known as the PTR, which activated the missile's onboard systems, was accidentally left in a live, post-launch position. The team in the command bunker spotted the error, but by then the rocket's electrical batteries were already powered up and live. Also active by now were the membranes on the fuel and oxidiser lines of the second stage; the parts of the self-igniting propellant were just one valve away from the engine's combustion chamber.

A nervous Chief Designer Yangel wanted a cigarette break. For safety reasons, Major General A G Mrykin advised him to do so in a bunker around seventy metres away. A G Iosifiyan, a leading engineer in electrical systems, argued that they needed to delay the launch one more time.

But there was no time to lose. Sergei Khrushchev (son of Nikita) wrote in his memoirs about the incident, and recalled a witness in the command bunker overhearing the exchange:

'So should I move PTR to zero?'

'Go ahead.'

As the PTR switch was moved to 'zero setting', it passed the 'on' setting and caused the rocket to ignite while still on the ground. At around 6.45 pm local time, half-an-hour before launch, the second-stage engine fired up, and flames burst from the fuel tank below, causing a massive explosion and a vast fireball. There were huge numbers of casualties among the hundreds of people in the area, many of whom were engulfed in flames, some instantly transformed to cinders, while others were overcome by poisonous gas. People scrambled to escape from the inferno but were held up by the security fence and by tar melting underfoot.

It took two hours to put the fire out. Forty-nine survivors were taken to intensive care, but sixteen of them were to die from their injuries. Most of the bodies were burned beyond recognition. There was no sign of Marshal of Artillery Nedelin, who had been present at the test site.

Khrushchev ordered Leonid Brezhnev to set off with no delay for Tyuratam to investigate. Brezhnev announced that there would be no further punishment as 'All guilty had been punished already.' The conclusion to the matter was that management of the testing had been overconfident about the safety of the vehicle, and insufficient work and analysis had been carried out beforehand.

The commission's final report read: 'The direct cause of the accident was the shortcomings in the design of the control system, which allowed unscheduled operation of the EPK V-08 valve controlling the ignition of the main engine of the second stage during pre-launch processing. This problem was not discovered during all previous tests.'

Russian sources refused to reveal the exact death toll in the R-16 accident. The secret document remained in the classified archive of the Central Committee of the Communist Party, only to be revealed after the fall of the Soviet Union. The official announcement made on 28 October 1960 was that all, including Marshal Nedelin, had perished in a plane crash. We now know that Nedelin and at least 125 other rocket personnel were killed in the R-16 blast. He was buried underneath the Kremlin wall. Others were buried in a common grave in the territory of Baikonur or shipped home. A memorial for the victims of Tyuratam was erected three years after the accident.

To this day, the Nedelin Incident stands as a grim warning. As one of the NASA officials present at the 2005 presentation pointed out, 'Thank God they didn't have a warhead on the missile that day.'

The USA, however, should not feel too smug about the Nedelin explosion – at least that Soviet catastrophe was a nuclear near-miss, a 'might have been'. The tale of the USAF's exploding Bomarc missile in New Jersey in 1960 posed a direct nuclear threat – it had a live nuclear warhead.

The Boeing Corporation has this to say about the history of its Bomarc missile (its name a portmanteau of 'Boeing' and an acronym of the University of Michigan Aeronautical Research Center):

> The supersonic Bomarc missiles (IM-99A and IM-99B) were the world's first long-range anti-aircraft missiles, and the first missiles that Boeing mass-produced ... Between 1957 and 1964, Boeing built 570 production missiles ... and another 130 for various tests. The missiles were housed on a constant combat-ready basis in individual launch shelters in remote areas ... On Oct. 1, 1972, the last Bomarc was retired from service.
>
> Michael Lombardi, 'Reach for the Sky',
> *Boeing Frontiers,* June 2007

The first models of the long-range, surface-to-air missile were armed with 10-kiloton nuclear warheads that were supposed to knock Soviet bombers out of the sky with a blast half as powerful as that of the atomic bomb dropped on Hiroshima. The missiles flew at altitudes of up to 70,000 feet (21,340 metres) with speeds approaching 3,000 miles per hour (4,830 kilometres per hour).

Ten Bomarc bases were set up across the United States and Canada. The first opened at Fort Dix in New Jersey under the command of the 46th Air Defense Missile Squadron at McGuire Air Force Base. The base had 56 missiles, each stored under its own concrete shelter with a movable roof. Every 14-metre-long rocket, called a 'pilotless interceptor', was kept on its launcher for quick firing.

Hardly had it been accepted into service when something went wrong. On 7 June 1960, a Bomarc nuclear-tipped missile ruptured its high-pressure helium tank, exploded and its liquid-fuel tank caught fire. A sentry nineteen kilometres east of the

McGuire airfield heard 'a dull thud' from a missile silo and saw the orange-yellow cloud rise into the air. Civilians reported sirens going off and emergency warnings in the neighbourhood. In the initial confusion, the state police thought a nuclear warhead had exploded. Although the flames were extinguished within an hour, airmen poured water on the smouldering rocket for fifteen hours, sending plutonium steam drifting downwind. While the missile's explosives didn't detonate, the heat melted the warhead, releasing plutonium, which the fire crews then spread around. Officials in Philadelphia later ordered panic tests for radioactive fallout in the air or water.

Bertram Gratz, an army reservist at Fort Dix for advanced infantry training, and squad leader, was returning with his squad from the local range; as they passed the missile base, 'all hell broke loose,' he later remembered. 'We saw this puff of black smoke come up . . . There were sirens going off all over the place.'

Gratz had read the air force's publicity about the new weapon and immediately worried that his men might be in danger. 'This thing, I believe, had a bursting radius of 1,000 yards [900 metres],' he recalled. 'The fireball would have engulfed us if we stayed where we were. I said, "Drop the guns and let's get the hell out of here."' They ran to the protection of a sand berm and listened to local radio reports concerning an alert at McGuire Air Force Base. The story was picked up in the United States and around the world. A story in New Jersey's *Trenton Evening Times* on 8 June 1960 carried the following headline: 'A-Missile Blaze Scare Sets Off Probe Demands'.

This was, however, in the middle of the Cold War, and the East German radio station Deutschlandsender (DLS) broadcast the embarrassing news to the world. The Soviet propaganda machine moved quickly on 9 June to follow up the Bomarc story and spread the news.

Clumsy Excuses
Moscow, Soviet European Service in English, June 9,
1960, 2130 GMT—L
(Vetrov commentary)

The world public is deeply concerned at the latest proof of how great the danger of an atomic disaster is in our time. And indeed a densely populated area near the McGuire base in New Jersey was on the verge of a terrible disaster on Tuesday, June 7, when something went wrong in the nuclear warhead of a Bomarc missile and the tail light in the explosive device burst into flames. As we all know, it isn't only U.S. territory where U.S. rockets with nuclear warheads are stockpiled, and U.S. bombers carrying atomic weapons are not only flying over U.S. territory.

And on 11 June, the Soviets rubbed salt into the wound by alleging that panicked citizens had been fleeing New York.

The US authorities tried to draw attention away from the story rather awkwardly with some classic disinformation, attributing the cause of the fire to a bursting helium gas bottle. But by then the cat was out of the bag.

The USAF allocated $6 million to recover the site at Fort Dix. The area was hastily cleaned up and the air force capped the contaminated soil with concrete and asphalt – effectively sweeping the radioactivity under the carpet. Plans to cart away 7,650 cubic metres of soil, concrete and steel were put on hold because, not unreasonably, surrounding communities did not want radioactive waste shipped through their neighbourhoods.

The fire became a footnote in the history of the Cold War. The base closed to the public in 1972, but an estimated 300 grams of plutonium from the melted warhead still remains in the sandy soil, entombed in asphalt and concrete – a radioactive relic of the Cold War and just one of the many toxic hot spots from the era that dot the USA. It was one of the worst publicly acknowledged nuclear accidents up to that time.

In 1985, New Jersey Governor Tom Kean voiced concern after learning about the incident. Although the fire was no secret, the nature of the contamination had been concealed until 1973. Most of the radioactive soil and debris material has now been removed and shipped to a disposal site in Utah.

Of particular concern was the fact that the base sits on top of the Kirkwood-Cohansey Aquifer, although federal officials have stated tests show that the plutonium has not affected the groundwater, which is fifteen metres below the surface in that area. Yet the military are still taking no chances. 'They test our water every year to make sure it doesn't get into our wells,' revealed George Mostrangeli, who lives three kilometres away. 'We don't drink the water out here anyway . . . The standards and theory concerning radiation was that you could just wash it away.'

The Bomarc accident makes the point that live nuclear warheads are inherently dangerous. Even if they don't detonate in an accident and cause a nuclear explosion, the possibility of radioactive contamination is always present.

An arguably more serious accident happened when a warhead fell off a missile inside its silo, Lima-02, near Vale in South Dakota. Very few people are aware of the incident, although during the Cold War the plains of western South Dakota were dotted with 150 USAF missile silos housing hundreds of Minuteman ICBMs, each loaded with 9-megaton multiple warheads that could strike the Soviet Union in around thirty minutes. The threat of nuclear war was terrifying, but the locals were kept unaware of a nuclear accident that could have demolished everything within 180 square kilometres (70 square miles) and caused thousands of deaths.

On 5 December 1965, a cold winter's night, twenty-year-old

weapons specialist Bob Hicks was in his barracks. The phone rang. It was the chief of his missile maintenance team: an incident had occurred at an underground silo, would Hicks take a look? 'The warhead is no longer on top of the missile,' said the chief, adding, unnecessarily, 'Be careful.'

When he arrived at the site, Hicks heard that during maintenance work on a missile, an airman inadvertently caused a short circuit with his metal screwdriver, which in turn had created a small explosion. The missile's nose cone, which contained the thermonuclear warhead, had been blasted off, plunging nearly twenty-five metres to the bottom of the narrow silo, finally coming to a rest on its side on the concrete floor. Luckily, the cone hadn't damaged the missile enough to set it off and the warhead did not appear to be damaged. The military believed that there was no immediate threat of a nuclear explosion, so only a *potential* broken arrow was declared. A radiation-monitoring team went into the silo but found no trace of a radiation leak, so Hicks volunteered to check out the warhead.

He climbed down the shaft and into the equipment room that encircled the upper part of the underground silo. There he inserted three rods to render the electrical connections safe and make the rocket's stages incapable of firing. Having attended to the missile, it was time to go deeper and resolve the issue of the fallen warhead. After Hicks had gone down and confirmed the lie of the nose cone, he reported back to the surface.

He suggested that a net could be lowered and the cone with its warhead could be rolled into the net. The net could then be hoisted up on a cable by a crane. After an argument with an officer (who thought that this could not be done), Hicks went back down the silo. Two cargo nets were lowered and he wrapped them round the cone, then hooked it securely to a lifting cable. Then came the delicate process of raising the cone up the silo, in the small amount of space available between the wall of the silo and the missile; if the nose cone had struck the missile, catastrophe would have ensued. After a long, slow lift, the cone finally reached the surface and was whisked away to Lackland Air Force Base in San Antonio to be taken apart. The accident report indicates that, the next day, the remainder of the ICBM was drawn from its tube like a bad tooth.

That same month, Hicks was awarded an Air Force Commendation Medal for his courage. 'By his personal courage and willingness to risk his life when necessary in the performance of dangerous duties,' the citation read, 'Airman Hicks has reflected credit upon himself and the United States Air Force.'

South Dakota's missile silos were decommissioned following the signing of the Strategic Arms Reduction Treaty by the United States and the Soviet Union in 1991. The silos now sit empty and overgrown, visited by curious tourists in the grassy wastes of the Black Hills.

But accidents continue to happen. In 2014, three airmen were conducting maintenance on a Minuteman III missile at a silo in Colorado when an accident caused $1.8 million worth of damage to the missile – roughly the same amount, taking inflation into account, as the 1964 accident in South Dakota. According to a brief statement from the US Air Force, it seems that a team was trying to troubleshoot the missile after it failed a diagnostic test and had become 'non-operational'. Ultimately, the accident would likely have been categorised as a 'bent spear' event, the code used by the military for impaired weapons.

The most chilling part of the story is the USAF's reluctant admission that the missile was partially damaged during the maintenance check because the maintenance chief 'did not correctly adhere to technical guidance' and 'lacked the necessary proficiency level' to understand that what was being done to find the problem. That bland confession is perhaps the scariest bit of all. Who wants badly trained fumble-fingers fiddling with nuclear missiles?

None of the accidents suffered by the US nuclear-weapons programme has ever caused a nuclear detonation. That there was not a detonation or a major leak of radioactivity at Lima-02 that day in 1964 is thanks to the bravery of a young airman, Bob Hicks.

These samples of accidents are but the tip of the iceberg. When we look at the overall US record of nuclear weapons mishaps on the ground, or try to unravel the Russians' records, and then add in the catalogue of accidents suffered in the UK, France, India and Pakistan, it merely reinforces the danger of nuclear missiles. Accidents happen.

A couple of incidents that happened in the UK demonstrate this inevitability. In 1976, a WE177 nuclear bomb fell off a workstand at RAF Honington in Suffolk, while being loaded on to a plane. In 1988, another WE177 was dented after it fell from a parked bomber at RAF Marham in Norfolk. Such mishaps are common and prove the point that nuclear weapons are always accidents waiting to happen.

On 8 August 2019, at Nyonoksa, near Archangel, the Russians were testing a new hypersonic, nuclear-powered cruise missile. Something went wrong. Russian media reported that a cloud of toxic gases had swept across a Russian town after a mysterious explosion was produced by radioactive isotopes released at the Nyonoksa testing range. A Norwegian nuclear expert told *The Barents Observer* that the isotopes of strontium, barium and lanthanum were evidence that it 'was a nuclear reactor that exploded'. Five military and civilian specialists were killed and up to six were injured. A local village was evacuated. Although Russia's explanations for what happened have varied, US intelligence officials suspect that the cause was a failed weapons test of the Burevestnik nuclear-powered cruise missile, a superweapon NATO calls the SSC-X-9 Skyfall. The radiation still contaminates the area and the seabed.

CHAPTER 13

RADIATION IS REALLY BAD FOR YOUR HEALTH

'Like taxes, radioactivity has long been with us and in increasing amounts; it is not to be hated and feared, but accepted and controlled. Radiation is dangerous, let there be no mistake about that [. . .] Consider radiation as something to be treated with respect.'
**– PROFESSOR RALPH EUGENE LAPP,
THE MANHATTAN PROJECT**

Splitting the atom and harnessing its power has been one of mankind's greatest achievements. Nuclear energy appears to be one of the most efficient forms available to us today, and modern radiation therapy allows us to fight cancer and other diseases with new curative tools.

But as we saw in chapter 2, the ruthless rays of radioactivity can pose a potentially deadly threat to humans and many other species as well. The radiation deaths after the explosions at Hiroshima and Nagasaki took the world by surprise, but there can be no excuse for the death toll brought about by radiation at Chernobyl, the victims of which will suffer for at least two generations. By 1986, the USSR knew all about radiation. We tend to deal with the ghastly horrors of Chernobyl and Fukushima by believing that as long as we don't live cheek by jowl with a nuclear power plant, there's no real need to worry. That is not true, as the separate chapters on those two nuclear catastrophes will illustrate, because the effects of nuclear radiation are rather more widespread. When dealing with nuclear radiation, we cannot afford for a second to let our guard down or relax our vigilance. Not only is it silent, invisible and deadly, but also it does tend to hang around...

There are now potential nuclear radiation dangers everywhere in the developed world, including unexpected quarters such as buildings, the world of fashion and even in what many would expect to be a safe haven – a hospital.

In 1985, Atomic Energy of Canada Limited (AECL) produced the Therac-25, a top-of-the-range radiotherapy machine. Its primary use is to kill off cancers. Like all radiation therapy machines, the Therac-25 worked by firing an intense, narrow beam of radiation at a body part afflicted with cancer. A linear accelerator produces beams of high-energy radiation that can be aimed precisely at the tumour. The treatment works by killing off cancer cells or keeping them from growing and dividing. It works well, but radiation is dangerous. Any overdose can burn tissue, damage cells and leave necrotic, open, weeping sores. Doctors specialising in radiation therapy have determined the optimum doses for specific types of cancer that maximise effectiveness and minimise any harm to healthy tissues. Radiation treatment has a high success rate and is considered a valuable weapon in the fight against cancer – providing that the dosage is correct (radiation therapy is based, after all, on the radium that killed Marie Curie).

Six AECL machines were used in Canada and five in the USA. The Therac-25 used a new beam-control system that relied on a flattening filter, which dissipated the radioactive beam over a larger area and weakened it. AECL's previous models, the Therac-6 and Therac-20, incorporated a physical safety feature: a hardware lock with a manual switch that prevented the beam being fired unless the filter was in position. In the Therac-25, this had been replaced by a computerised lock.

Unfortunately, the Therac-25 software had a bug. The filter system needed eight seconds to absorb its new instructions and could be compromised if the operator typed in the locking command too quickly. If this happened, the radioactive beam could fire without the filter in place, hitting the patient with a radiation overdose. Operators became accustomed to the machines occasionally freezing up and ignored or overrode the problem. That resulted in some patients receiving unfiltered radiation at far too high a dosage, without anyone realising what was going on. Such doses could be lethal.

In June 1985, Katie Yarborough attended the Kennestone Regional Oncology Center in Marietta, Georgia, for her twelfth cancer treatment. Having had a lump successfully removed from her left breast some months previously, she had been undergoing radiation treatment in the adjacent lymph nodes to prevent any return of the disease. Yarborough was used to the painless treatment lasting only a few seconds, but on this occasion she felt an intense heat and complained to the technician operating the machine that he was burning her. He responded that it simply wasn't possible. Two weeks later, Yarborough returned to the centre and showed medical physicist Tim Still a red mark the size of a penny on her chest. There was also a larger pink circle of skin high on the left side of her back.

Still was alarmed, telling Yarborough and her doctor, 'That looks like the exit mark made by an electron beam.' The pulse of energy had gone right through the unfortunate patient like a death-ray bullet. Still later estimated that she had probably received between 15,000 and 20,000 rads on that penny-sized space. Katie Yarborough's injury developed into a hole that, despite the best efforts of surgeons, could not be covered by skin grafts. She lost the use of her left arm and was in permanent pain, but she continued with her life as best she could. Her eventual death, over ten years later, was through injuries sustained in a car accident. Other patients died as a direct result of radiation over-exposure within months.

Dr Still discussed the potential problem with the machine with professional colleagues and found the manufacturer's response disturbing. 'I got this intimidating phone call from AECL,' he remembered. 'I got told that this kind of talk was libel unless I had proof and that I'd better stop.'

The machines were involved in six accidents between 1985 and 1987. A later report observed, 'Therac-25 operators had become accustomed to frequent malfunctions that had no untoward consequences for the patient.' To make matters worse, instead of registering that it had fired an intense dose of radiation, the computer screen read 'no dose'.

On 21 March 1986, an oilfield worker named Ray Cox was being irradiated for the ninth time at the East Texas Cancer Center in Tyler, Texas, for a tumour that had been removed from his back. At one point, feeling intense heat and pain, he jumped off the table. Cox complained immediately and the incident was logged. The machine had hit him with 15,000 rads and he died a few months later. Because the unfiltered beam had been delivering potentially lethal doses, only three of six Therac-25 patients who had reported burning and shocks survived.

The problem was finally identified by a member of the hospital staff, physicist Fritz Hager, and his technician, who had operated the machine during Cox's accident. They played with the various settings of the computerised machine to discover why it inexplicably delivered a massive overdose sometimes. Eventually they came upon a 'Malfunction 54' error message, caused by the computer's refusal to accept new commands and display them on the screen. The new screen data would look correct to the technician but, inside the computer, the software would already have encoded the old information for delivery. Put simply, the Therac-25 could be a killer.

That evening, Hager telephoned AECL to let them know that the Therac-25 could become a lethal death ray. Despite the inevitable wrangling and legal blame game that followed, there could only be one outcome. In 1991, AECL received a formal ban on all its medical equipment as a result of the Therac-25 malfunction.

This medical radiation glitch is by no means unique. Another example comes from Bialystok in Poland. In February 2001, at the city's oncology centre, the NEPTUN 10P radio-therapy machine lost power during a routine treatment for breast cancer. Power cuts were commonplace in the clinic. The session resumed once the power was up and running again and the machine had been checked over. A further four patients, all with post-surgical breast cancer, were treated with the same machine.

The staff became suspicious after the individuals reported burning and itching after treatment. Upon investigation, it became apparent that the machine was giving off significantly higher doses of radiation than necessary – and for no obvious reason. Further inspections revealed that an electrical part within the safety system had become damaged, which was impairing the machine's dose regulation and monitoring system. It transpired that although the clinic had made the correct checks during their restart of the machine after a power cut, checking the dose of radiation was not in the manual. A subsequent report stated:

> Machine shut-offs due to power cuts had happened many times in the past. The AC mains voltage in the hospital

area was quite unstable and as many as two power cuts a day had occurred occasionally. The experience of the radiation technologist with *previous power cuts* indicated that, after resuming operation, the machine performed normally...

Fortunately, none of the five patients had received lethal doses, though all required skin grafts for their burns. The hospital received a fine, and the consultant in charge was tried for criminal negligence, but was acquitted of any wrongdoing.

Not all radiation accidental exposures are the direct consequence of human or mechanical error. Sometimes they arise from plain laziness and negligence. During the 1980s, in the town of Kramatorsk in Ukraine, several of the inhabitants of one particular apartment block began to die of mysterious ailments. In 1981, a year after her family had moved in, a teenage girl died in one of the apartments; the following year, her sixteen-year-old brother passed away too, followed by their mother. All had succumbed to leukaemia. A new family settled and within two years lost their teenage son – also to leukaemia. The local doctors blamed poor heredity and bad ecology, but the father of the deceased boy (who had had 'the energy of a bulldozer', in the original Russian) took up an investigation into the causes, and began his own research into the likely cause of his son's death.

He suspected radiation as the most likely explanation, but why was it so localised? How had it happened? Where had it come from? He contacted a health physicist who immediately recognised the common symptoms of radiation poisoning: why were the white cell counts so low on all the victims? He took a Geiger counter and began to investigate the building. What he found horrified him. Sure enough, something in the building was pumping out dangerous levels of radioactivity.

Eventually they located the source. In the late 1970s, in

one of the nearby granite quarries, the work team had lost a capsule with radioactive caesium from a gauge used to measure radioisotope levels. In true Communist style, the loss had never been reported. No one was willing to tell the boss or take the blame.

Somehow, the radioactive source had ended up mixed with the gravel and aggregate used in 1980 to construct apartments at Gvardeytsiv Kantemirovtsiv Street in Kramatorsk. It lurked somewhere in a concrete wall, emitting lethal levels of radioactivity. In nine years, six residents of the building died from leukaemia and seventeen more absorbed lethal doses of radiation.

Eventually, the small capsule containing highly radioactive caesium-137 was found deep inside an inner concrete wall of Apartment 85, Building 7; it was pumping out deadly gamma rays with an exposure rate well over the safety limit for humans. In the words of the final report, 'If you're listing places where you wouldn't want radioactive material to wind up, lodged in a concrete block directly above a child's bed ...' would probably be high on the list. With its dangerous half-life of thirty years, caesium-137 behaves just like potassium, distributing itself throughout the human body, but in caesium's case, causing sickness and death because it attacks living cells. In today's brave new nuclear world, you don't need to wait for a nuclear war to become a nuclear victim.

Others have unwittingly brought disaster on themselves. Sometimes – like the 'radium girls' discussed in chapter 2 – you can fall victim to the horrors of radiation without realising it. On Christmas Eve 2001, a call came in to the International Atomic Energy Agency (IAEA) Emergency Response Centre in Vienna from Georgia in the Caucasus. In the mountains near the border with Abkhazia, a trio of woodcutters had discovered two strange metal objects; they were warm enough to have melted the snow around them. The night was freezing and the snow thick in the forest, so the delighted woodsmen took them back to their makeshift hut for warmth.

During the night, however, they soon became seriously ill, with dizziness and headaches; they were also struck by vomiting bouts. All three were sent to a hospital in the capital, Tbilisi, where their skin became afflicted with huge burns, though they were eventually to recover. From the symptoms, physicians suspected they had suffered radiation poisoning. The Georgian government called in the IAEA.

The mysterious 'heaters' turned out to be abandoned Soviet radioisotope thermoelectric generators (RTGs), stripped of their protective lead encasing for reasons unknown. This rendered them so toxic that the IAEA employed twenty-five people to move them into drums lined with lead; each team member had a mere forty seconds' worth of exposure to the radiation.

RTGs transform heat from radioactive strontium-90 isotopes into electricity; they power satellites and space probes, among other devices. The IAEA quickly identified the two mysterious objects as the highly radioactive remains of a former Soviet communications system. The Soviets had scattered strontium-90 devices throughout the far-flung corners of the USSR, as batteries and power sources for isolated communications relay stations in remote areas. According to Professor James Clay Moltz, a specialist in, inter alia, international affairs, Russian foreign policy, international space strategies and international nuclear weapons and policies at the Monterey Institute of International Studies, 'I'd say there are hundreds of these things out there . . . It's something of a concern, and it's really irresponsible of the Russian and post-Soviet nuclear authorities not to have gathered them in. It's a hazard to the population.'

John Holdren, a physicist at Harvard University's Program on Science, Technology and Public Policy, agrees. He stressed that strontium-90 is 'certainly very dangerous stuff . . . and the industrial and medical equipment often contains material that is even more radioactive and poses disposal problems.'

The three woodcutters were lucky and survived. But their story raises the question of just how many more of these lethal nuclear batteries lie abandoned and scattered somewhere deep in the former Soviet Union. No one knows.

If the Georgian woodcutters were victims of bad luck when they accidentally 'embraced' radiation, they at least did it for the sound practical reasons of keeping warm on a freezing night. On the other side of the world in 2013 something almost as dangerous happened – but as a fashion statement. The French saying *'il faut souffrir pour être belle'* ('it's necessary to suffer to be beautiful') would have been horribly realised in January 2013, had it not been for alert US customs inspectors on duty with a Geiger counter. They discovered that the studs in a consignment of belts for fashion retailer ASOS had been made from highly radioactive cobalt-60, a nuclear isotope more commonly used in radiotherapy. Containing up to eight hundred metal studs, the belts could have been dangerous if worn for more than a month or so of everyday wear. The headline from a report on the story on slate.com summed it up: 'These Belts Are Hot (Because They're Made from Radioactive Scrap Metal)'.

It was alleged that an Indian company called Haq International had supplied ASOS with the shipment of 641 radioactive belts, although Haq International disputed the claims. Nevertheless, the company said that it had cancelled a £64,000 order with the Indian supplier. Neither ASOS nor the supplier were allowed to independently test the belts, which were being held by US customs officials.

ASOS eventually recalled the products, which had been sold in fourteen different countries, and commissioned a report. It makes for worrying reading:

Unfortunately, this incident is quite a common occurrence, since manufacturers in India and the Far East use scrap metal for their products. When scavengers look for scrap metal, they sometimes come across 'orphaned' radioactive sources, which have been forgotten or abandoned by authorities instead of being properly and safely disposed of. If a radioactive source and other raw materials are melted together at the same time, the radioactivity becomes trapped in the resulting metal alloy.

ASOS stated that it had subsequently contacted all of the customers who purchased the product, and it appears that they were eventually recalled.

The problem is not new and is much more widespread than we realise. One thousand La-Z-Boy recliners had to be recalled in 1998 because they contained radioactive metal components. In fact, between 2003 and 2008, the Department of Homeland Security in the USA turned back in excess of 120 contaminated shipments containing, among other things, cutlery and tools.

Scarce and high-value items are a source of profit. Many sensitive materials often end up on the scrapheap. Scrap is often at the heart of the matter. The workers in India made fashion belts for ASOS from components sourced from other vendors, who often got their metal from scrapyards. Junkyards in the developing world regularly buy their raw materials from poor, often desperate people, scrabbling through waste to feed their families.

In 2000, a disused radiotherapy unit containing a cobalt-60 source was stored at an unsecured location in Bangkok, Thailand; part of it was then accidentally sold to scrap collectors, who transported it to a junkyard in the province of Samut Prakan. Unaware of the dangers, an employee there dismantled the device and extracted the source, which remained unprotected for several days. Ten people, including the scrap collectors and workers at the junkyard, were exposed to high levels of radiation and became ill. Three of the junkyard workers subsequently died as a result of their exposure. Afterwards, the source was safely recovered by the Thai authorities

Remarkably, officials have noticed that the junkyards surrounding Chernobyl, of all places, are mysteriously shrinking. Abandoned cars have been relieved of their engine blocks and other large chunks of metal are missing. There is a serious danger here. All those abandoned cars are positively throbbing with induced radioactivity.

One blatant criminal act involving radiation was a robbery in Mexico that went badly wrong. In December 2013 at the village of Tepojaco in Hidalgo, a radioactive cobalt-60 teletherapy source en route to a waste-storage centre was hijacked. According to a statement from the IAEA, the cobalt-60 was packed within protective casing and safe when it was loaded on to the truck, but could become toxic if uncased or damaged in some way.

Unsurprisingly, when the vehicle was tracked down, near the town of Hueypoxtla, the casing was opened, although the source was found not far away. In due course, the Mexican Health Secretary confirmed that six men had been admitted to the General Hospital of Pachuca, and that the suspects would be turned over to investigators for questioning. Hidalgo's state health department later reported that the sextet, aged between sixteen and thirty-eight, were reported to be stable – and under arrest.

One of the most notorious radiation accidents caused by theft happened in Brazil on 13 September 1987. The incident is considered to rank among the top ten most severe nuclear disasters to date, categorised as a Level-5 accident on the seven-stage International Nuclear Event Scale (INES) – in which each stage represents an incident of ten times greater magnitude than the previous stage. (For a more detailed description of the way this scale works, turn to chapter 19.) Thanks to the research of Meg Girli at Stanford University, we have a fairly clear idea about what happened. At the end of 1985, a private radiotherapy institute, the Institute Goiano de Radioterapia in Goiânia, Brazil, moved to new premises, taking with it a cobalt-60 teletherapy unit and leaving a caesium-137 tele-therapy unit behind in an unsecured building. Brazil being Brazil, two thieves promptly moved in and stole the radiation head of the machine unit, thinking they could sell it for scrap.

They took their booty home and tried to dismantle it, exposing the radioactive source capsule. It contained our old friend caesium-137, a radioactive isotope in the form of caesium chloride salt, which is highly soluble and readily dispersible. The thieves then sold the remnants of the source assembly to a junkyard owner. He noticed that the material glowed blue in the dark and, fascinated, invited people to come and have a look.

Over the next few days, friends and relatives visited to marvel at this phenomenon and even took some pieces home as glow-in-the-dark novelties. After about a week, some of them began to fall ill. The symptoms were not initially recognised as being related to irradiation but one of the people affected, who had become very sick, connected his illnesses with the source capsule and took his piece of caesium to the public health department in the city. Bureaucratic panic ensued. The contamination sites were swiftly identified and residents evacuated. The Brazilians recognised that they had a serious problem on their hands.

Twenty people were subsequently admitted to the Marcílio Dias Naval Hospital in Rio de Janeiro and another six patients to the Goiânia General Hospital. Four of the casualties died within weeks of their admission. The post-mortem examinations showed haemorrhagic and septic complications associated with the acute radiation syndrome. In total, some 112,000 persons were monitored, of whom 249 were seriously contaminated either internally or externally. With the widespread contamination involving caesium-137, experts from the United States and the Soviet Union travelled to Brazil to help in what was described in *The New York Times* as 'a radiation accident now proving to be the most serious of its kind in the Western Hemisphere'.

It became clear that the entire town must now have become contaminated by the release of the caesium. A massive clean-up operation was required. Seven main areas of severe contamin-ation were identified, with more than 67 square kilometres (26 square miles) of urban Goiânia at risk. Mass evacuations were ordered, eighty-five houses were identified as badly contaminated and the bulldozers moved in. Seven houses were demolished. Tonnes of soil were removed and the ground covered with concrete or fresh earth. A laboratory was set up in Goiânia for measuring the caesium content of soils, ground-water, drinking water, air and foodstuffs.

Once again radiation had shown its hidden potency. But radiation is ubiquitous. Every day we receive a belt of solar radiation from the sun. UV rays may give us a healthy-looking tan, but we are all now well aware that although sunlight is an essential prerequisite for life, it can threaten humans. Excessive exposure to the sun is known to be associated with increased risks of various skin cancers such as melanomas, cataracts and other eye diseases, as well as accelerated skin ageing.

Not all natural radiation, comes from above, however. Some of it is lurking in the ground. A good example turned up the Kwale region of southeastern Kenya. In the mid-1990s, vast deposits of titanium were discovered in this coastal region. Surveys showed that it was in an area that was naturally radioactive and that most of the south coastal area of Kenya underlain by radioactive rock formations. Mrima Hill in the Kwale district, in particular, registers as highly radioactive. Geophysicist Jayanti Patel, who had conducted a study of the area years before the prospectors arrived, advised local authorities of the inherent danger, telling them, in his own words, that 'Sedimentary rock from the hill should not be used for either building homes or road constructions.'

Levels of radioactivity emitted by Mrima Hill were multiple times higher than those considered acceptable. Among its lethal list of elements are uranium, potassium-40 and thorium. So many local people had died from unexplained causes that the hill had become notorious. Inevitably bowing to economic pressures, the government issued a contract to make repairs to a dirt road and open up the area by using materials from Mrima Hill, rather than using rock from another quarry nineteen kilometres away. As with most government schemes to save money, it went badly wrong. Part of the dirt road contained a dangerous level of the compound thorium-232. The upshot was that anyone who took part in the repair of the road, plus an estimated 25,000 people living nearby, was contaminated by radiation, most likely from dust. They had to have check-ups after radioactive materials were used for road repairs, according to Kenya's *Sunday Nation* newspaper.

In all, approximately 25,000 people had experienced radiation poisoning, albeit at a non-lethal level. As well as continuing to track their health, the Kenyan government took away 2,795 tons of the materials employed in road repairs, and made Mrima Hill a protected zone. Whether the clean-up was complete, or the affected victims were adequately cared for, remains open to question, given the lack of resources in the region.

But we cannot blame nature or misguided science for all the problems caused by radiation. Human fallibility also contributes. Perhaps one of the saddest cases was an incident in October 2011, at the Venerable Hospital of St Francis of Penance in Rio De Janeiro, Brazil. Seven-year-old Maria Eduarda was undergoing treatment for acute lymphoblastic leukaemia, which had been diagnosed the previous year. She'd already undergone chemotherapy and this was to be followed by a course of radiotherapy.

Much to their dismay, her parents noticed that her skin was coming out in burn marks after the radiotherapy sessions. Maria's doctor advised them that these often resulted from radiation treatment. But in due course, burns appeared on her head too and Maria began to display difficulty in speaking and walking. Eventually, it was discovered that she had frontal lobe necrosis – internal brain damage. She died in mid-2012.

This time the fault lay not in the machine but with the human operators, three of whom would subsequently be charged with manslaughter. The doctors and technicians had made a terrible mistake. They had incorrectly calculated the number of radiotherapy treatments Maria was meant to receive. Instead of measuring out the doses she was to receive over eight sessions, they had given her a full dose each time. The front of the little girl's brain had been burnt out with nuclear radiation.

CHAPTER 14

YOU CAN'T BRUSH NUCLEAR UNDER THE CARPET

'Irresponsibility on the part of the radioactive material's owners, usually a hospital, university or military, and the absence of regulation concerning radioactive waste, or a lack of enforcement of such regulations, have been significant factors in radiation exposures.'
– ATOMIC ENERGY AUTHORITY, GENEVA REPORT

The biggest, most enduring problem with nuclear, never mind accidents or human error, is that like Br'er Rabbit's Tar Baby, it just won't go away. Radiation is dangerous when it is discovered or created, and it may well remain dangerous forever.

Plutonium – very nasty and highly radioactive – has a frightening shelf life. The longest-lived is plutonium-244, with a half-life of 80.8 million years. Plutonium-242 has a half-life of 373,300 years, while the plutonium-239 used in nuclear bombs has a half-life of 'just' 24,110 years. You can hardly stuff nuclear waste under the mattress or sweep it under the carpet, so how do we best dispose of it?

The Hanford Site is a US decommissioned nuclear production complex in Washington State that dates back to World War II. Hanford had released thousands of litres of radioactive liquids into the wild. Many of the people living in the affected area received low doses of radiation. Washington, DC, kept quiet.

The Hanford weapons production reactors decommissioned at the end of the Cold War left a problem: 200,600 cubic metres of high-level radioactive waste, stored in almost 200 tanks; 708,000 cubic metres of solid radioactive waste; and vast areas of groundwater below the tank farms contaminated with technetium-99 and uranium.

Hanford is currently the most contaminated nuclear site in the United States. The locals have inherited a ticking time bomb of radioactive waste, and it won't go away. But this problem does not only affect the United States – it is a worldwide issue. A sample of nuclear incidents just from the UK highlights the extent of the international headache.

Most people have not heard of the Office for Nuclear Regulation (ONR), the UK's specialist nuclear watchdog, which was established in 2011, barely a month after the Fukushima disaster. Set up to regulate nuclear safety and security and to ensure that obligatory safeguards for the UK are met, the ONR is an independent statutory corporation reporting to the Department for Work and Pensions and working closely with such bodies as the Health and Safety Executive and the Department of Energy and Climate Change. While the number of formally declared nuclear safety incidents appears to have been stable for a decade, the overall rate of 'faults' recorded by the watchdog has doubled since 2010 to more than one a day. As the years pass, the nuclear problem, and particularly that of nuclear waste disposal, is building up. The role and responsibilities of the ONR are interesting because they highlight the contradiction at the heart of governmental control and regulation of nuclear affairs.

On the one hand, this body must stand guard over a large, potentially lethal and ageing estate, encompassing everything from nuclear submarine docks to secret warehouses where degraded fuel is stored in glass. On the other, in true Whitehall style, it is boxed in, forced by the UK Treasury rules to pay for itself commercially with 94 per cent of its budget coming from the very companies it was set up to regulate.

The ONR documented a total of 973 'anomalies' from 2012 to 2015. These were either unrated (meaning that they were not regarded as having had any significance, in the context of nuclear risk) or allocated an INES score of zero. Here are some of them, as highlighted in an article by Oliver Moody published in *The Times* on 27 December 2016.

- Four cases where tritium, a radioactive form of hydrogen, was found at elevated levels in local groundwater around the Dungeness B nuclear power station reactor in Kent.
- At least seventy safety incidents on the UK's main nuclear warhead base at Aldermaston, Berkshire, including the contamination of several workers and a power cut across the site.
- An accident where a vehicle carrying nuclear material on the M1 hit a lorry, and another where a transport vehicle flipped over, damaging two containers holding radioactive chemicals.
- Uranium 'sludge' and an unstable form of caesium left in bin bags at Springfields, a former power plant, and Amersham nuclear materials factory.
- At least a dozen leaks of radioactive substances and more than thirty fires at power stations, including an event where a control panel at the Sellafield site was burnt out.

Questions have been raised about these statistics and the ONR's assessments. Brief accounts of the incidents that were recorded in the three years up to March 2015, but were logged as being of no significance are controversial. Nuclear experts maintain that some of these basically unreported events were actually more serious than many of the accidents officially noted by the ONR as safety issues with an INES (International Nuclear Event Scale) score of one or higher.

The ONR's reports show that radiation alarms at Britain's ports and airports were set off on fifteen separate occasions (including four at Heathrow) by packages thought not to contain any radioactive material. They also mention that thirteen workers at various sites had worryingly high radiation counts, that a worker at the Devonport nuclear submarine base in Plymouth breathed in an unstable isotope of cobalt, and that a contractor at Harwell swallowed plutonium. Very few of these incidents made the press.

Sellafield, the renamed Windscale in Cumbria, has recorded 167 nuclear-related incidents. These include contamination of the surrounding ground, power cuts and unplanned shutdowns, and, alarmingly, a complete loss of cooling water around the reactor. Dr Paul Dorfman, Honorary Senior Research Associate at University College London's Energy Institute and founder and Chair of the Nuclear Consulting Group, warns, 'The seriousness and scale of these INES incidents... is bad news for nuclear safety in the UK.'

Benjamin Sovacool, Professor of Energy Policy at the University of Sussex, agreed with Dorfman. Quoted in *The Times* article of 27 December 2016, he remarked that the ONR is 'both cheerleader and policeman'. The ONR insists that its safety ratings are strictly in line with international guidelines and that there is no conflict of interest in its role:

We are robust in upholding the law and use our regulatory enforcement powers to hold the industry to account wherever necessary. The rating of nuclear safety 'events' is based on agreed international criteria and it is wrong to suggest that we would seek to 'downplay' these. The standards of safety we expect from the nuclear industry are extremely challenging and the majority of events are of very minor nuclear safety significance.

The question, however, remains. Does the ONR work perhaps a little too closely with the thirty-seven nuclear sites that it regulates, even asking them to draw up their own safety regulations? This may or may not be a good thing; the debate continues.

Contrast that, however, with the Russian whitewash over an incident that has been described as 'a cover-up of nuclear fallout worse than Chernobyl'. The scandal centred on the effects of nuclear bomb testing in the 1950s. During that decade, and into the early years of the 1960s, Semipalatinsk in Kazakhstan was the location for the world's highest count of nuclear bomb tests. In August 1956, fallout from Soviet nuclear weapons testing at Semipalatinsk swept over the nearby industrial city of Ust-Kamenogorsk, hospitalising more than six hundred people with radiation sickness. Moscow's desire to hush things up, concealing the effects of a nuclear disaster that had long-term consequences on its citizens' health, remained secret for more than half a century.

The story first went public in 2017, when Britain's prestigious *New Scientist* magazine got access to a hitherto secret Soviet report and revealed news of 'a nuclear disaster four times worse than Chernobyl in terms of the number of cases of acute radiation sickness'. The Kremlin had covered up radiation exposure and associated health impacts on civilians living close to a Kazakhstani testing site. The report was found in the archive of the Institute of Radiation Medicine and Ecology (IRME) in Semey, Kazakhstan, after the fall of the Iron Curtain. 'For many years, this has been a secret,' said the institute's director Kazbek Apsalikov, who discovered it.

As reported in *New Scientist*, it all started after a 1956 nuclear test. In 2002, Konstantin Gordeev at the Institute of Biophysics in Moscow published a map showing that, on 24 August 1956, a radioactive cloud travelled directly over Znamenka and Ust-Kamenogorsk, both downwind of the nuclear test range. The findings tally with previous reports of the path of the fallout clouds. A test on 12 August 1953 had sent another lethal cloud over Karaul, the effects of which would endanger public health for three years.

As a result of the Soviets' own scientific monitoring, Moscow set up a body to monitor public contamination, which built up a record of 100,000 people and their children whose health had been jeopardised by the tests. The facility was disguised as the 'Anti-Brucellosis Dispensary No. 4' so as to disguise its true purpose. Only in 1991 was the clinic's work declassified.

The Kremlin had good reason to keep quiet. In mid-September 1956, a month after the fallout cloud hit, 638 people were hospitalised with radiation poisoning – more than four times the number of cases admitted as diagnosed after the 1986 Chernobyl disaster. Worse, dose rates were up to 1.6 rads per hour, a hundred times what the International Commission on Radiological Protection recommends as safe.

The final report, outlining 'the results of a radiological study of Semipalatinsk region' and marked 'top secret', showed just how much Soviet scientists at the time knew about the human-health disaster in the late 1950s – and how desperately the Soviet authorities tried to hush things up. Moscow researchers made three special expeditions to the Ust-Kamenogorsk region; they discovered contamination not only in the food and ground there, but also in urban areas of eastern Kazakhstan. More worrying still, among other cases cited by locals were long-term genetic birth defects, cancer and thyroid problems.

The story slowly emerged after the collapse of the Soviet Union in 1991. The 'dispensary' became the IRME but, according to its former chief scientist, Boris Gusev, much of its archive was destroyed or removed to Moscow before the handover of the USSR to the Russian Federation. The report stands as a glaring indictment of Moscow's attempts to conceal its nuclear excesses and how the Communists tried to hide the truth from their own citizens, never mind the rest of the world.

The weapons testing site was officially closed in 1991, but these attempts to hide the effects of radioactivity at Semipalatinsk pale into insignificance compared to the Soviets' nuclear problems at Mayak, in the southern Ural Mountains, in the late 1950s. Although his NKVD spies had warned him of the Americans' nuclear progress, Stalin had shut down the Soviets' existing metallurgic research station and sent his scientists to work in the mining industry instead. The nuclear facility at Mayak started in 1945 as a panicky knee-jerk reaction to the Americans' atom bomb. It was only when the Soviet leaders saw the tremendous power of nuclear weapons, as demonstrated during the attack on Hiroshima and Nagasaki, that they realised how seriously behind they were in the nuclear arms race. The Great Leader knew that he had made a big mistake.

Accordingly, Mayak was built in just three years between 1945 and 1948. Lavrentiy Beria, the head of the NKVD secret police, no less, was put in charge of the USSR's top-secret nuclear programme. He drafted in over 40,000 prisoners from his gulags, plus German prisoners of war, to build the plant. In true Soviet style, Mayak was nowhere to be found on Soviet public maps but, set at the heart of a 250-square-kilometre (97-square-mile) exclusion zone in the Chelyabinsk Oblast (administrative district), it was to become Russia's largest nuclear reactor facility, sprawling over 90 square kilometres (35 square miles).

Scant regard was given to safety or how best to get rid of contaminated material. Radioactive waste from spent nuclear fuel was stored underground; when no more space was available, it was thrown into the sluggish river Techa. Downstream, its water was used by more than 100,000 people. Furthermore, contaminated water from the reactors was discharged straight back into Lake Kyzyltash, but water from that lake was also used to cool the same reactors. The whole area gradually became a contaminated radioactive zone.

One of the first recorded accidents occurred in 1953, but went unnoticed until a worker developed radiation sickness so severe that both legs had to be amputated. But the so-called 'Kyshtym disaster' – the secret Mayak facility often being referred to by the name of the nearest-known town – was on a different scale altogether. On 29 September 1957, the cooling system of one of the waste-storage tanks failed and the tank exploded spontaneously with a force of some seventy tonnes of TNT. A plume of radioactive fallout rose into the sky, reaching a height of around 3,050 metres.

In the afternoon, residents in the Chelyabinsk district noticed a strange phenomenon in the sky, similar to the aurora borealis. Naturally, because of the lock-down secrecy surrounding Mayak, the Soviet authorities did not tell the locals what had happened or warn them of the risk of fallout. Over the next few days, the radioactive cloud drifted slowly northeast for several hundred kilometres, contaminating an estimated 15,000 to 20,000 square kilometres and putting the lives of 270,000 people at risk. Only when villagers started to become ill with the classic symptoms of radiation sickness did the Soviet authorities panic and take action. A week after the incident, they began a hasty evacuation of the affected area. Over the next two years, at least ten thousand people were secretly evacuated and twenty-two villages were abandoned.

Inevitably, vague rumours of a Soviet nuclear catastrophe

Above left: Marie Curie, 1867–1934 – a Polish-born naturalised Frenchwoman, she was a pioneer in research into radioactivity, and is the first and only woman to win two Nobel Prizes, and in two disciplines, Physics and Chemistry. *(© Alamy)*

Above right: New Zealand-born Ernest Rutherford, later first Baron Rutherford of Nelson, 1871–1937 – 'the father of nuclear physics'. *(© Alamy)*

Below left: The German-born theoretical physicist Albert Einstein, 1879–1955. Although a pacifist, in 1939 he wrote to US President Franklin D Roosevelt, warning of the dangers posed by Nazi Germany and the need for the USA to investigate nuclear weapons. *(© Alamy)*

Below right: The Danish physicist Niels Bohr, who won the Nobel Prize in Physics in 1922 for his contribution to the understanding of atomic structure and quantum theory. *(© Alamy)*

Left: Harry S Truman (1884–1972), who succeeded Roosevelt as US President on the latter's death in April 1945. It was under Truman that the atomic bombs were dropped on Hiroshima and Nagasaki in August 1945, a decision he took to save American lives. *(© Alamy)*

Right: Lieutenant-General Leslie Groves (1896–1970), the US Army Corps of Engineers officer who directed the Manhattan Project; he also oversaw the building of the Pentagon. *(© Alamy)*

Left: US Admiral Hyman G. Rickover (1900–86), 'the father of the nuclear navy'. Born in Poland to a Polish Jewish family, he came with them to the USA in 1906, to escape Russian anti-Semitic pogroms. *(© Alamy)*

Left: 'Little Boy', the atomic bomb dropped on the Japanese city of Hiroshima on 6 August 1945. *(© Alamy)*

Right: 'Fat Man' – the weapon that devastated Nagasaki three days later. Six days after that, Japan surrendered unconditionally, so ending the Second World War.
(© Alamy)

Below: A Japanese survivor of the bombing of Hiroshima, showing radiation burns suffered as a result of the blast. *(© Alamy)*

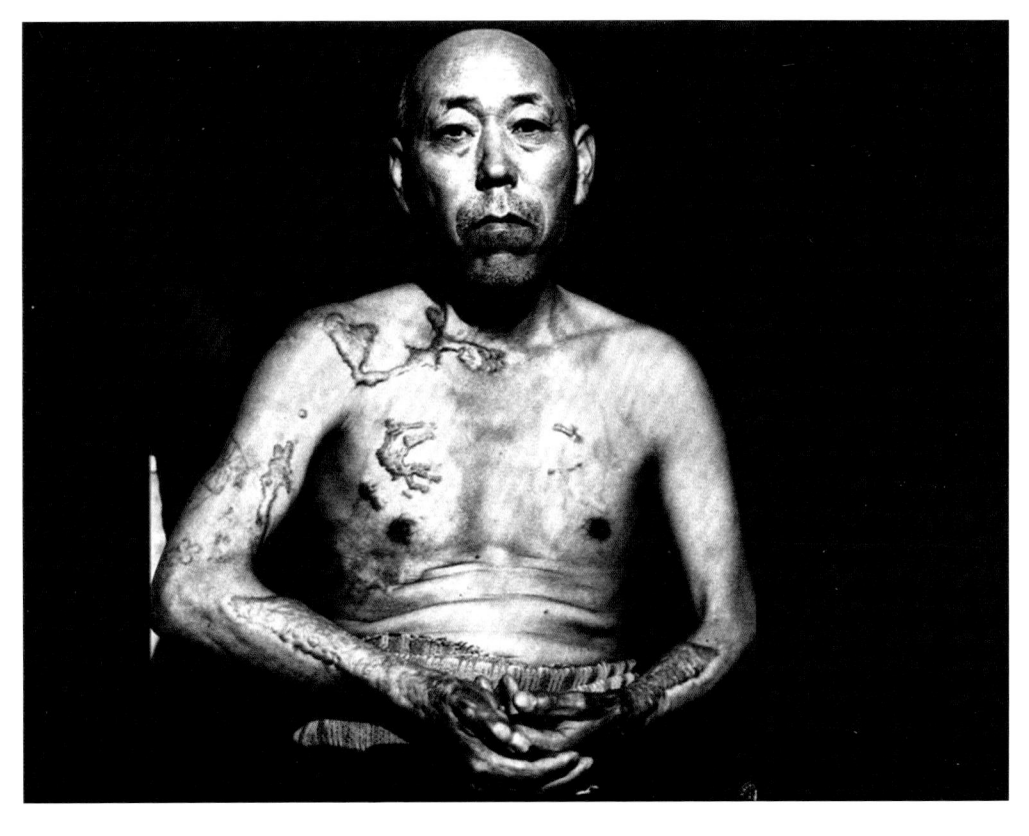

Right: A US hydrogen-bomb test at Bikini Atoll in the Marshall Islands. After the inhabitants had been evacuated, the US tested 23 nuclear bombs on the islands and lagoon that make up the atoll, out of a total of 67 tests in the Marshall Islands. *(© Alamy)*

Left: A Convair B-36 Peacemaker, the USA's first dedicated nuclear bomber, and the largest mass-produced piston-engined aircraft ever built. On 22 May 1957, a USAF B-36 accidentally dropped a 20-tonne H-bomb near Kirtland Air Force Base, New Mexico, although luckily the weapon was not armed and only the conventional trigger charge detonated. *(© Alamy)*

Right: A Boeing B-47 Stratojet, in service with the USAF from 1951 until 1977. Incidents involving B-47s included one aircraft crashing into nuclear bunkers at RAF Lakenheath in Suffolk, England, in July 1956, killing its crew, although none of the nuclear weapons or their trigger charges detonated.

(© Alamy)

Right: *K-19*, Soviet Russia's first nuclear-powered submarine, launched in 1959. In fact, the vessel suffered so many defects and accidents that its crew nicknamed it 'Hiroshima'. This RAF reconnaissance photo of *K-19* was taken in March 1972 after the vessel had been crippled by a fire 800 miles east of Newfoundland and had to be towed home; thirty of her crew died as a result of the incident.
(© Getty Images)

Left: A Boeing CIM-10 Bomarc long-range surface-to-air missile, in service with the USAF from 1959 until 1972. On 7 June 1960, at McGuire Air Force Base, New Jersey, a Bomarc caught fire in its silo, destroying the installation and releasing plutonium, although the nuclear warhead did not detonate. *(© Alamy)*

Right: A USAF Titan II intercontinental ballistic missile (ICBM), in service from 1962 to 1987. At Little Rock Air Force Base, Arkansas, on 18 September 1980, an accidentally dropped wrench ruptured a Titan missile's fuel tank, and the volatile propellant exploded some hours later, killing an airman and destroying the silo. Although its nuclear warhead was blown about thirty yards, neither the conventional nor main charges detonated. *(© Alamy)*

Left: The Greenpeace protest vessel *Rainbow Warrior* in Port of Auckland, New Zealand, after she was bombed and sunk by two agents of the French secret service on 10 July 1985. The ship was to be used to protest against French nuclear tests in the Pacific; one of her crew, a photographer, died in the explosion.

(© Getty Images)

Right: Alexander Litivinenko, a former agent of Russian intelligence who fled to Britain and became an outspoken critic of Putin's Russia, in his hospital bed in London in November 2006, not long after he had been poisoned with polonium-210, almost certainly at the behest of the Russian state. He died on 23 November.

(© Getty Images)

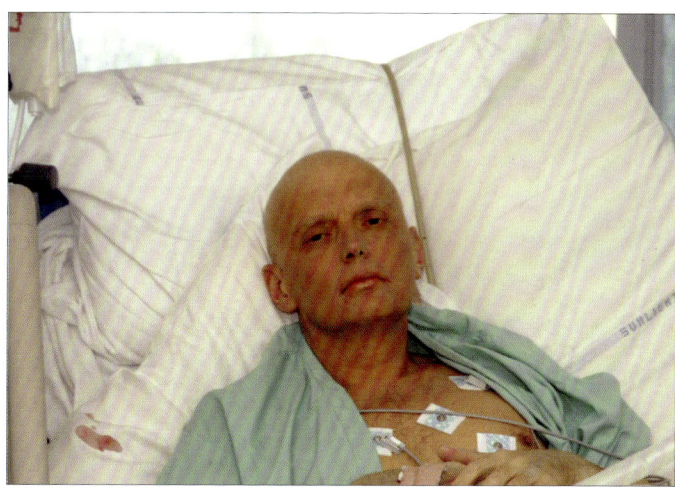

Left: A Boeing B-52 Stratofortress nuclear bomber, introduced in 1955 and still in service with the USAF. On 17 January 1966, a B-52 collided with its refuelling tanker and crashed, dropping three hydrogen bombs on land near Palomares in south-eastern Spain, and a fourth into the sea. None of the weapons' main charges detonated, but two of the conventional trigger charges did.

(© Alamy)

Left: The Windscale (now Sellafield) nuclear plant at Seascale in Cumbria, which began generating in 1950. The Windscale reactor fire of 10 October 1957 burned for three days and spread radioactive fallout across the UK and Europe. It is the UK's worst nuclear accident to date, rated at 5 – out of a maximum 7 – on the International Nuclear Events Scale (INES). *(© Getty Images)*

Right: The nuclear plant at Three Mile Island, Pennsylvania. The partial meltdown of Reactor 2 and radiation leak on 28 March 1979 was also rated 5 on the INES.
(© Alamy)

Left: The nuclear and radiation accident at the Chernobyl plant, in what is now Ukraine, on 26 April 1986 is rated 7 – the highest – on the INES. A quarter of a million people were evacuated and the area is still closed off; it has been estimated that there will be 25,000–100,000 cancer deaths in the fifty years following the incident. *(© Alamy)*

Right: Like Chernobyl, the disaster at the Fukushima-Daiichi nuclear power plant in Japan, about 150 miles (240km) north-north-east of Tokyo, on 11 March 2011 is rated 7 on the INES. For all its reputation as a world leader in technology, Japan has a poor nuclear safety record.
(© Alamy)

'The man who saved the world' – Lieutenant-Colonel Stanislav Petrov (1939–2017), formerly of Soviet Russia's Air Defence Forces, who on 26 September 1983 correctly interpreted a radar trace of five incoming US ICBMs as a false alarm, thereby almost certainly averting a nuclear holocaust. Regarded as a hero in the West once the story got out, he was treated shabbily by the Soviet military. *(© Alamy)*

began to leak out. As early as 1958, a Danish newspaper ran a story alleging that a major accident had been a factor in the Soviets' decision to unilaterally suspend nuclear tests. Other papers picked up on it. True to form, the Soviet government indignantly denounced the reports as unfounded, provocative Western propaganda.

A more complete story began to emerge in 1976, when Zhores Medvedev, a dissident Soviet biologist living in Britain, published a series of articles about the disaster in *New Scientist*. His allegations were confirmed by the eye-witness testimony of another Soviet scientist, Lev Tumerman, who later fled to Israel. In a letter to *The Times* and the *Jerusalem Post*, he noted:

> About 100 kilometres from Sverdlovsk, a highway sign warned drivers not to stop for the next 20 or 30 kilometres and to drive through at maximum speed. On both sides of the road, as far as one could see, the land was 'dead': no villages, no towns, only the chimneys of destroyed houses, no cultivated fields or pastures, no herds, no people... nothing.

Later, to researchers at the Los Alamos National Laboratory in New Mexico, he expressed his astonishment at the barren landscape near Sverdlovsk, with entire villages destroyed and no sign of life or crops.

Many scientists at the time doubted Medvedev's account as exaggerated allegations and 'hearsay'. Even the Soviets couldn't hush up a disaster of this magnitude, surely? However, US intelligence sources found some corroboration for his story, reporting that a 'vast contaminated area had been identified northeast of the Mayak plant'.

Concrete evidence of the true scale of the disaster only emerged after the break-up of the Soviet Union in 1991. Intelligence officers and scientists alike were taken aback by the sheer magnitude of the disaster. Rated as a Level-6 incident on the INES, the Kyshtym catastrophe is acknowledged as the third most serious nuclear accident recorded after Chernobyl and Fukushima.

Paranoiac Soviet secrecy hid the truth. At this juncture, it's worth noting the safety record of the Mayak plant over the years, with over thirty incidents reported since 1953. These have ranged from the Kyshtym disaster of September 1957, discussed above, in which an explosion caused the release of a strontium-90 radioactive cloud as well as caesium-137, to a criticality accident in December 1968 when a plutonium solution was poured into a cylinder with 'dangerous geometry' leading to one fatality and one other case of radiation sickness, eventually requiring the amputation of both legs and an arm. As late as September 1998, equipment and pipelines of the first circuit at Mayak were contaminated, after the output of its P-2 reactor ('Lyudmila') was increased, leading to the permitted power level surging to over 10 per cent and the failure of three channels of the fuel rod seal, contaminating equipment and pipelines.

The fall of the USSR did not lead to much of an improvement in the twenty-first century in this litany of hushed-up nuclear accidents. When handling procedures of liquid radioactive waste led to waste disposal in open water in 2003, Mayak lost its operating licence for a time. Yet more accidents subsequently occurred, involving radioactive pulp, spillage of radioactive liquid (both 2007), and the release of alpha emitter in 2008, which caused a repair worker to lose a finger. Even in September 2017, a mysterious radioactive increase across Europe, notably in the level of radioactive pollution in the village of Argayash, was traced back to the nearby plant of Mayak.

And it has not just been the Russians. How many people have heard of America's Savannah River spillages? At least thirty times between 1957 and 1985, a nuclear weapons plant near Aiken in South Carolina experienced what one scientist subsequently termed 'reactor incidents of greatest significance', including leakage of radioactivity and a meltdown of nuclear fuel. Not one of these was reported to local residents or to the public generally.

The story did not come to light until a US congressional hearing exposed it in 1988. The Savannah River plant was operated by E I du Pont de Nemours and Company. DuPont was accused of covering up the truth about the accidents. The company immediately issued a denial, pointing out that it had routinely reported all such events to the Department of Energy. Later, the Department of Energy admitted that it had tried to keep the news secret.

Others have been guilty of trying cover up nuclear mishaps, too. Governmental claims to have a monopoly of, and tight control of, all matters nuclear is as full of holes as a Swiss cheese. The Yorkshire saying 'where there's muck there's brass' applies to nuclear just as to any other commodity. There is money to be made in dirty business. The introduction of more rigorous environmental legislation in the 1980s made illegal waste dumping a lucrative affair. Organised crime quickly saw the potential for rich pickings in matters nuclear, especially when it came to hiding governments' embarrassing nuclear waste – for a price, naturally. Ever quick to spot a chance to make a fast buck, the Italian 'Ndrangheta moved in.

The 'Ndrangheta is a particularly nasty, tight-knit Calabrian organised-crime syndicate (The name is old Calabrian from a Greek word meaning 'heroes'.) The gang has been around since the late 1700s and is particularly difficult to penetrate because it is linked primarily by blood ties and has been run for generations as 'the family business'. An American investigator described the organisation as 'a Mafia within the Mafia'.

According to a 2013 'Threat Assessment on Italian Organised

Crime' from Europol and the Guardia di Finanza, the 'Ndrangheta is one of the richest and most powerful organised-crime groups in the world. SISMI, the Italian Secret Intelligence Service, claimed that the organisation had been involved in radioactive waste dumping since the 1980s.

For a time around the turn of the century, pirates controlled the approaches to Aden, at the eastern end of the Red Sea and across a narrow strait from Africa, but a greater crime is still being perpetrated by multinational companies using the African mainland as a toxic-waste dumping ground. In 1997, in the Italian magazine *Famiglia Cristiana*, Greenpeace published a landmark investigation into the dumping, which showed that in the late 1980s, Swiss and Italian companies acted as brokers for the transportation of hazardous waste from Europe to dumps in Somalia. Subsequent research has also shown that the company employed physically to ship the waste was wholly owned by the Somali government.

An investigation into the murder of the Italian journalist Ilaria Alpi in Somalia as early as 1994 revealed that she was killed because she was about to blow the whistle on the Somalian guns-for-waste trade and the dumping-at-sea racket. Greenpeace managed to get the European Green Party to raise a question in the European Parliament about 'the dumping of toxic waste from German, French and Italian nuclear power plants and hospitals in Somalia. An Italian investigation (Italy having been the former colonial power in Somalia) later concluded that around 35 million tonnes of waste had been exported to Somalia for $6.6 billion, leading the environmental group Legambiente to assert that Somalia's inland waste dumps are 'among the largest in the world'.

To make things worse, the Boxing Day Tsunami of 2004 washed up corroded drums of unidentifiable ooze on to Somalian beaches; suddenly local villagers began to die of unexplained illnesses. A subsequent UN Development Programme (UNDP) report pointed the finger at illegal disposal, concluding that the 'dumping of toxic and harmful waste is rampant in the sea, on the shores and in the hinterland'.

A year later, a Somali NGO conducted its own survey. It identified fifteen containers of 'confirmed nuclear and chemical wastes' in eight coastal areas and reported a sudden rash of chronic and acute illnesses suffered by local Somalis. These included severe birth defects, such as the absence of limbs, and widespread cancers. One local doctor said he had treated more cases of cancer in a single year than he had in his entire professional career before the tsunami.

Bringing those responsible for the dumping to justice may be impossible, however. The originating country is responsible for disposing of its medical and nuclear waste, as well as for its retrieval if it is disposed of illegally. With many of the damning containers unmarked and much of the paperwork probably long since lost or destroyed, legal action is impracticable. Yet deeper investigation revealed an even nastier truth: the whole European radioactive and toxic dumping racket was a 'Ndrangheta front.

The story exploded when Francesco Fonti, a former member of 'Ndrangheta, exposed the conspiracy in the magazine *L'Espresso* in 2005 and turned state's evidence. He openly accused the 'Ndrangheta of illegally sinking at least thirty ships loaded with toxic waste, much of it radioactive, in return for payments. The waste was dumped by scuttling clapped-out ships off Calabria and then moved to Somalia and the Far East. Fonti's revelations led to widespread investigations into such waste-disposal rackets.

Taiwanese environmental groups were furious and demanded that their government launch their own investigation into the allegations and conduct tests to determine whether the dumping of waste had had an impact on Taiwan's environment and the condition of the ocean. The 'Ndrangheta struck back in typical ruthless style – two investigative reporters from Italian state broadcasting network RAI, sniffing round similar criminal activities off East Africa, were mysteriously killed in Somalia.

But deeper investigation reveals that the illegal dumping of nuclear waste has long been big business in Italy. Extremely toxic waste has been offloaded around Naples for some time. As early as 2004, the British medical journal *The Lancet Oncology* described the area around Acerra as a 'triangle of death', where sheep with two heads were born. 'We are talking about millions of tons of nuclear waste,' a former leading light in the Casalesi clan, which was largely responsible for Naples's toxic woes, warned government investigators in 1997. 'I also know that trucks came from Germany carrying nuclear waste.' Some 11.6 million tons of waste are illegally disposed of annually in Italy, a practice that earned the underworld in excess of €16 billion in 2012, according to Italian environmental group Legambiente.

And no one dares to drink the local water. The reasons are carefully listed in a $30-million study commissioned by the US Navy's 6th Fleet in 2011 and which was brought to the attention of Italians via an article in *L'Espresso*. Its alarming title? 'Drink Naples and Die'. The Americans found that water from 92 per cent of the private wells sampled outside the base posed an 'unacceptable health risk'. In 5 per cent of the samples, uranium levels were found to be 'unacceptably high'.

Further west in Giugliano, the poor live in shacks and caravans at the foot of what is probably Europe's nastiest landfill, stuffed with, among other things, toxic nuclear sludge and dioxin. According to a geological study, the poisons will continue to 'contaminate dozens of square kilometres of land and everyone who lives there'. Antonio Marfella, from the Italian Cancer Research Institute in Naples, offers other sober findings. Tumours have increased by 47 per cent among men in the province of Naples, within the past three decades. The region of Campania now has the highest infertility rate in Italy and also leads in cases of severe autism.

In the end it is national governments that must accept responsibility. We cannot blame organised crime for manufacturing deadly nuclear isotopes. Crooks dump the waste where no one is looking in return for dodgy payments and no questions asked. But the issue is just too big and too dangerous an issue to be supressed by bureaucrats desperate to pretend that there is no real problem. The sad truth is that the worst cases have really been all about bribery and corruption and then saving officials embarrassment and hiding the truth from the suspicious taxpaying public.

No case makes this better for the complicity of bureaucrats everywhere than the infamous saga of California's 1959 partial meltdown of the Sodium Reactor Experiment (SRE). In the 1940s, the US government sought out a safe place to set up a rocket engine and nuclear reactor testing facility for new equipment that was considered too dangerous to experiment with near populated areas. Santa Susana in the hills overlooking the San Fernando Valley, about forty kilometres from downtown Los Angeles, seemed ideal. The Santa Susana Field Laboratory was established as a remote facility.

Over the years, ten different nuclear reactors operated at Santa Susana, as well as plutonium and uranium fuel fabrication facilities and a special 'hot lab' where highly irradiated fuel from all over the USA was shipped for examination. Tens of thousands of rocket engine tests were also conducted at the site, including those for the Apollo Moon landing programme and ICBMs.

Since the lab was first built, the population in the Los Angeles area has grown to the extent that more than 150,000 people live within eight kilometres of the site and about 500,000 people live within double that radius. But over the years there were several reactor accidents, the most notorious in 1959. The SRE was an experimental nuclear reactor that operated at the site from 1957 to 1964. For some unknown reason there was a power surge in July 1959 in the reactor core, which operators were barely able to shut down. Although they could not identify the cause of the problem, the reactor was started up again and ran for ten more days, in the face of rising radiation levels and clear indications of fuel damage. When the reactor was finally shut down, thirteen of the reactor's forty-three fuel elements had partially melted, and the facility had deliberately released radioactive gas into the atmosphere to stop an explosion on the site.

There followed a decades-long cover-up of the incident by the US Department of Energy. Five weeks after the accident, Atomics International, the company running the site for the US government, reluctantly issued a press release stating that 'a parted fuel element' had been observed, 'but there were no indications of unsafe operating conditions' and that there had been 'no releases of radioactivity'.

It became apparent that radioactivity had been venting into the atmosphere for weeks. The SRE's runaway accident was the first commercial power plant in the world to experience a core meltdown. Denial was standard operating procedure in the entire nuclear industry. Unfortunately, the terriers of the Californian environment lobby had got their teeth into the government and its nuclear lies. It turned out that Santa Susana environmental practices were downright dangerous. Radioactive and chemical wastes were routinely burned in open-air pits with barrels of waste torched by firing flares to set them alight, sending plumes of toxic smoke drifting far beyond the site.

The final straw came after the destructive Woolsey wildfire, which swept through the California Hills. The conflagration ignited on 8 November 2018 and burned 96,949 acres of land. Despite denials from the US federal authorities, evidence proved that the blaze had actually started on the Santa Susana nuclear site. The Department of Toxic Substances Control reassured the citizens of California that none of their measurements for radioactivity or toxic chemicals had found any impurities whatsoever at the location.

This was an extraordinary statement, because some years before the US Environmental Protection Agency had spent $40 million and several years testing part of the Santa Susana site, identifying hundreds of locations with elevated radioactivity and numerous toxic hot spots. Other agencies spent tens of millions of dollars uncovering thousands of locations around the Santa Susana Field Laboratory area with chemical contamination. For state and federal agencies now to assert they couldn't detect any such pollution at Santa Susana leads to one of two conclusions, both deeply troubling. Either their equipment was too insensitive to detect the contaminant that they had already identified was present – and thus their conclusions were worthless. Or the government was lying.

The locals had no doubt about the answer. Lawyers quickly uncovered damning admissions by South California Edison (SCE) that electrical infrastructure on site was involved with an 'event' at around the time the fire started. Initial investigations showed that an SCE substation power line failed near Chatsworth inside the Santa Susana Field Laboratory. Since then, California lawyers have recommended dealing with each individual claim rather than a 'class action'. That will cost. Once again, the nuclear chickens had come home to roost.

As ever, it proves that you just can't brush nuclear under the carpet, but this litany of hushed-up serious nuclear accidents is worthy of deeper study. The events reveal incompetence, ignorance, a casual attitude to safety and government cover-ups on a grand scale. One surprising example reveals exactly how governments deal with their nuclear accidents. In 1957, the British government perpetrated a cover-up so huge that it made the Russians look like amateurs in the official state secrets game...

CHAPTER 15

BRITAIN'S DIRTY SECRET: THE WINDSCALE FIRE

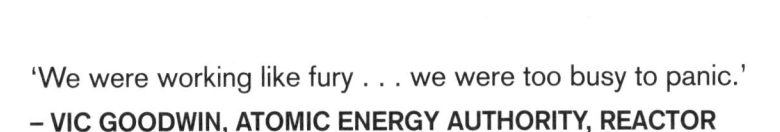

'We were working like fury . . . we were too busy to panic.'
– VIC GOODWIN, ATOMIC ENERGY AUTHORITY, REACTOR PHYSICIST AT WINDSCALE

The incident was billed as Britain's worst-ever nuclear accident, but time and new information have revealed an even darker and more tortuous story. The great Windscale fire turns out to have some very murky aspects indeed, not least the British establishment's complicity in what was a major whitewash. Britain is in no position to lecture the Russians or the Americans over nuclear cover-ups and lies, unless those lectures are to teach lessons on how to hoodwink the public.

The story really starts with Britain bankrupted by a second disastrous world war and desperately trying to retain its status as a great global power, even though its gold reserves and economic assets had been picked clean by its voracious wartime ally, the United States. At the end of World War II, Prime Minister Winston Churchill was hoping to continue the 'special relationship' between the United Kingdom and the

United States, based on their wartime intelligence sharing and nuclear cooperation.

Churchill genuinely thought that the British contribution to the development of the atom bomb and the whole-hearted incorporation of Britain's ground-breaking nuclear Tube Alloys into the Los Alamos atomic facility and programme qualified the United Kingdom as an equal partner. The hard-nosed men in Washington did not agree. Having broken the British Empire economically, the US State Department was now determined to cut Britain loose to sink or swim in the harsh post-war world. The USA was no friend of the British Empire.

'You helped, but we did it,' observed one US nuclear historian and, in 1946, the US made it a capital offence to divulge nuclear secrets to any other country, friend or foe. Suddenly, spy-ridden, exhausted, impoverished London was on its own. Washington was determined to keep the monopoly of the atom bomb to itself.

Britain's new post-war Labour government, led by Clement Attlee, was outraged. Ministers and civil servants advised him that the country could not afford to compete with the USA. However, Ernest Bevin, as bellicose and patriotic a foreign secretary as any hard-line Tory imperialist, disagreed. He was determined to preserve Britain's world role. 'We have got to have this thing whatever it costs and we've got to have the bloody Union Jack flying on top of it,' he declared. Faced with a large and still fully mobilised Red Army now firmly camped in Eastern Europe and threatening the rest of the continent, Labour's principal aim was to maintain Britain's position on the world stage by the development and use of high-technology weaponry to deter any aggressor. The policy had another objective as well: to convince America that Britain was its nuclear ally. The UK had to reconstruct the enormously expensive experiments of the US nuclear programme. It did, however, have some formidable and experienced individuals to call upon. In particular: John Cockcroft, who set up the Harwell Atomic Energy Research Establishment; Christopher Hinton, who was tasked with building a nuclear reactor at Windscale as fast as possible; and Sir William Penney, who had been heavily involved at Los Alamos. This triumvirate was given the task of coordinating Britain's top-secret nuclear programme from scratch and assembling an atomic bomb at a facility in Aldermaston near Oxford. It was predicted that the Soviet Union would have their own atomic bomb by 1952, so that became the deadline imposed by the British government for the Windscale project. Hinton was worried by this timetable, pointing out the project's experimental nature and concerns about reactor safety. Despite these doubts, the work went ahead.

In the USA, the first reactor had been built in an uninhabited desert region with a 48-kilometre-long escape road to minimise risk. In Britain, only the Scottish isles offered such isolation and they were actually too remote and inaccessible. A compromise was reached. Whitehall decided that building the new reactor at Windscale near the village of Seascale on the Cumbrian coast in the northwest of England was an acceptable risk. The site had plenty of cooling water from Wastwater lake and was remote from the population in case of any nuclear accidents.

Work began at Windscale in 1947, before the research work at Harwell had been completed, and Seascale quickly drew some of the finest scientific minds in the country. Most of the newcomers were clever young physicists, hailed in the newspapers as the 'atom men' who would bring in a new age of scientific and technological achievement to create better lives for the people of Britain. Contemporary books and magazines celebrated the marvels of radioactivity with pictures of irradiated fruit and vegetables staying fresh for months.

Windscale soon had to deal with a major problem, however – how to dispose of the enormous heat generated by the reactor in the middle of a sparsely populated area. They solved this by installing a cooling system using huge fans to drive air up through the reactor and out through an enormous, 120-metre chimney. A year into the design and construction of the plant, Terence Price, one of Cockcroft's team at Harwell, asked a crucial question: 'What would happen if a uranium rod at Windscale caught fire?' A burning fuel rod could produce radioactive particles, which the cooling system would eject through its huge chimney. Price suggested that filters be fitted on the chimney, but his idea was dismissed, with his colleagues sceptical that any filter system could cope with the volume of up to two tonnes of material per hour rising through the chimney. Cockcroft, however, agreed with Price and ordered massive concrete filters to be built on top of the chimney. They became known derisively as 'Cockcroft's Folly'.

The Windscale project was finally completed in 1951, becoming a serious producer of plutonium, although not sufficient amounts to manufacture an atomic bomb. Worse, the Russians were now working on a new thermonuclear hydrogen bomb. The new Tory prime minister, Harold Macmillan, decided that a much bigger version of an atom bomb, called 'Orange Herald', was needed, something that would have almost as much destructive power as an H-bomb. Whitehall ordered Windscale production to be accelerated. The new weapon required massive supplies of plutonium and tritium, so the demand on Windscale was increased by 500 per cent. The scientists and engineers raised their collective eyebrows: this was easier said than done.

The only way to increase production was to allow the uranium powering the reactor to become even hotter, by removing all the aluminium fins used to dissipate heat from the reactor rods. This made them more likely to overheat. John Harris, a scientist employed at Windscale, recalled that while some scientists thought it 'great that we were getting enough plutonium', there were others who considered overheating the uranium an unacceptable risk and warned of the potential dangers. Nevertheless, all half million cooling fins were removed. Hinton, the chief scientist at Windscale, was to resign in protest within weeks of the Orange Herald test.

The plant was now under enormous pressure, faced with producing tritium (a radioactive isotope of hydrogen, essential for the British bomb) by further modifying the fuel rods in a now ageing reactor that had only been designed to produce plutonium and, on top of that, ensuring that it ran hotter to keep up with demand. Despite the risks, magnesium-lithium isotope cartridges were added to the fuel rods and in February 1952 the first small billets of plutonium were delivered (in the boot of a taxi) to the Aldermaston weapons factory.

Things were not going smoothly up at Windscale. Under these new pressures the reactor core began behaving unpredictably. Since modification, the rods had been producing unexpected bursts of energy, leading to sudden heating and the danger of fire. The scientists and engineers on site were carrying out controlled releases of the stored energy known as 'Wigner releases'. This involved allowing the core to heat up for a limited period, in the expectation that the nuclear energy accumulating in the rods could then be converted to heat and be blasted up the chimney.

More problems occurred when some of the rods became so hot they melted and fused on to the back of the reactor. The operators were reduced to dislodging them manually using metal poles. Later, men had to use shovels to remove the toxic slag and were exposed to dangerous levels of radioactivity in the process. It was around this time that a research scientist at Windscale, called Frank Lesley, began, on his own initiative, to take readings of radioactivity levels around Seascale. They were abnormally high. He reported them, was duly reprimanded by officialdom for his pains and, although the government was informed, Whitehall gave orders that the matter should remain top secret, concealed even from those making decisions about the reactor's future.

Events were coming together to invite an accident. There is now abundant evidence that the Windscale fire – the first major recorded blaze in any nuclear facility – happened because of pressure from the British government in their reckless drive to produce bomb-making material for nuclear weapons. Its real cause was the corner-cutting and abandoning of safety measures in order to meet these government targets – as was the case in the Soviet Union.

The potential hazard was doubled in October 1956, when the Calder Hall nuclear power station opened near Windscale. It was hailed as the first nuclear power station in the world, and the boast was that it would produce free electricity that was 'too cheap to meter'. Its first task was actually secretly to boost Windscale's production of nuclear materials needed to meet the increased demands of Britain's expanded nuclear bomb programme. A year later, on 10 October 1957, a serious fire broke out at Windscale.

Three days earlier, workers monitoring the temperature gauges had noticed that the temperature of the reactor core was rising. They ordered a Wigner release to try and cool it down, but the release did not have the expected result. The rods did not cool down; they continued to heat up. The primitive reactor temperature measuring instrumentation in the control room failed to record this. The operations room staff mistakenly thought the reactor was cooling down too much. They ordered an extra boost of heating. That was a big mistake.

The second Wigner release was ordered and the air cooling increased to take away any excessive heat. That didn't work either. The core continued to overheat and increasingly high levels of radioactivity were detected. The operators diagnosed a burst fuel cartridge; the truth was that one of the cartridges had actually caught fire. The increased airflow from the second

Wigner release spread the fire to other fuel rods and fire then engulfed the reactor. The blaze caused radioactive material to be ejected into the air.

The reactor that burned was one of two air-cooled, graphite-moderated natural uranium reactors at the site used for production of plutonium. The fire became an inferno, with blue flames leaping violently out of the reactor's rear face, blowing heavy concrete cooling vents out as they did so. Radioactivity and red-hot flames roared up the chimney. Despite the Cockcroft filters on the top of the chimney working full blast, they could only restrain a tiny amount of the burning material.

Down in the control room, they had no emergency protocols for dealing with a major fire. Windscale deputy general manager Tom Tuohy stepped in to lead the efforts that day and was tasked with trying to extinguish the unprecedented blaze. Hughes and another control-room operator went to the face of the reactor wearing protective equipment and were astonished at what they found. 'We saw to our complete horror, four channels of fuel glowing bright cherry red,' he recounted.

'Mankind had not faced a situation like this,' commented another witness. 'We had to play it by ear.'

The burning radioactive waste poured into the outside air. There were two major releases of radiation, one on 10 October and a second the following day, when water was used on the fire. An estimated 20,000 curies of radioactive iodine escaped, along with other deadly isotopes such as plutonium, caesium and the highly toxic polonium. The poison was carried away on the wind. It was a disaster with enormous ecological implications. The local residents of Seascale were completely unaware of what was going on. No official warning of the accident or the danger was issued.

On Friday, 11 October, several thousand employees at Windscale and Calder Hall were sent home. Guards were posted at the gates of both plants, while inside Windscale engineers and scientists tried desperate measures to deal with a radioactive fire raging at 1,300°C, completely out of control, roaring up the chimney to spew out red-hot radioactive particles. A cloud of toxic fallout formed, which then began drifting towards cities in northern England.

For the operators, the first priority was obviously to extinguish the blaze, the temperature of which was now increasing by 20°C with every passing minute. In an effort to create a fire break, teams working in relay used poles from scaffolding to push together piles of fuel cans. Attempting to stifle the conflagration by pumping in carbon dioxide gas proved counterproductive – generating more oxygen, which actually boosted the fire. Eventually, the idea emerged to flood the overheating reactor with water, but this presented its own risks: should acetylene gas or explosive hydrogen be produced, the reactor could blow apart. Nevertheless, after consulting with scientists on site, the works' general manager, Mr H G Davey, decided to try it. 'We had to ensure a real torrent of water,' he told a reporter. 'Too little water could have resulted in the release of hydrogen, so we pumped water in at the rate of 1,000 gallons [about 4,000 litres] a minute and kept on pumping it.' Overnight, the facility's own firefighters fought relentlessly to attempt to bring the temperature of the pile down.

This vast deluge of water on to the overheated nuclear reactor went on continuously for three days before, starved of oxygen, the fire died down and the reactor stabilised. The temperature inside the pile slowly began to fall. By Sunday night, it had cooled down to such an extent that only a small volume of water was necessary to keep it at a safe level.

On the afternoon of 11 October, Whitehall issued a statement:

> At 4:30 pm yesterday it was discovered that some of the uranium cartridges in the centre of Pile One at

Windscale, which was shut down at the time for routine isolation of the uranium and for maintenance work, had become over-heated to the point of red heat. The combustion is being held. The staffs are now injecting water on it from above and the temperature has started to fall. Some oxidation of the uranium has occurred.

The greater part of this has been returned by filters in the Windscale chimneys. A small amount has been distributed over the works site and in some areas works personnel, as a precautionary measure, have been instructed to stay under cover.

Health physics personnel are carrying out a check on both the site and in the surrounding district to ensure that any increase in the amount of radioactivity may be known. There is no evidence of there being any hazard to the public.

The type of accident that has occurred could only occur in an air-cooled, open circuit pile and could not occur at Calder Hall or any of the power stations now under construction by the Electricity Authority. At this stage it is not possible to give the cause of the accident. It is likely the Pile will be out of operation for some months. Further reports will be issued.

This was a classic of British understatement and obfuscation and a bare-faced lie. As the news flashed around the world, reporters sniffed a major story and swooped. A spate of inquiries flooded in to local newspapers, and national reporters bombarded Whitehall and Windscale with embarrassing questions. In the best traditions of Whitehall bureaucracy, the civil servants prevaricated, downplayed the real dangers and, where necessary, lied. *The West Cumberland News* of 12 October best sums up the official story:

> Rumours of a fire in one of the piles at Windscale works atomic energy factory swept through West Cumberland yesterday morning and they gained in strength when hundreds of men not employed on maintenance and safety work returned to their homes by bus and train.
>
> A strict security guard was thrown round the two sites and officials refused to answer telephone inquiries from the press. A news reporter who toured the district found no signs of panic. Some people living within a mile [two kilometres] of Sellafield were not even aware of anything untoward...
>
> Two hundred yards [182 metres] from the pile I saw septuagenarian Mrs Stanley, whose family had been Lords of the Manor for over 600 years, calmly planting out wallflowers in the garden in front of her cottage. I asked if she was at all worried by what was happening at the atomic factory. She replied, 'No, why should I be? If there was anything to worry about they would have told me by now. In any case my family have been here for over 600 years and I am staying here.'
>
> Just across the road I saw farmer John Bateman, he told me he was not worried and in fact did not know much about what had happened. Down at the village school I asked headmaster Joseph Tracey if any arrangements had been made for evacuating the children if there was any increase in radioactivity in the district. He said all that he knew about it was what he had heard on the radio and no evacuation had been considered.

The next step was to deal with the fallout. The radioactive hot plume had been carried on the wind for hundreds of kilometres. It was detected in Belgium, Denmark, Germany and Norway and may have been carried further east. Closer to home it posed a serious challenge to food supplies.

Following the fire, environmental measurements were taken to determine the radiological impact of the disaster. The results indicated that the highest concentrations of emitted fission products were iodine-131 and caesium-137. Of the two, the greater focus was on the iodine-131 as the most dangerous hazard from the fire.

Disturbing reports came from the monitoring vans that were touring the district testing vegetation and air. The ground was contaminated for an area of some 36 square kilometres (14 square miles) around Windscale. The grass was radioactive. The toxic iodine would be ingested and passed into the digestive system of cows, contaminating their milk. On the Saturday night, a ban was placed on the use of milk from all dairy herds within the affected area. Police and Milk Marketing Board officials toured the area from midnight onwards, waking up farmers and instructing them neither to sell their milk nor consume it themselves, nor to allow livestock to drink it.

Radioactivity affecting the food supply became the main concern. Authorities began to monitor the downwind external radiation levels. The highest reading obtained was on the Calder Farm road, where it was ten times greater than the ICRP level for continuous lifetime breathing. There were several competing estimates for the level of caesium-137 that had been released in the form of particulates, and air filters were used to sample the data. Aside from caesium-137 and iodine-131, the scientists warned that dangerous levels of tritium, polonium-210 and plutonium-239 had also been released.

Wide-scale sampling in the downwind area began and resulted in 1,140,000 litres of milk, over an area of 500–750 square kilometres (200–300 square miles) from roughly 600 herds of cattle, being condemned and thrown away. Fortunately, given the relatively short half-life of the iodine – two to three weeks – the situation was only threatening for a month.

It transpired that the radioactive footprint was much bigger than first admitted. Milk deliveries from twelve milk producers within a three-kilometre radius of Windscale had been stopped that first night. However, throughout Sunday and Monday, milk samples were collected from farms at increasing distances from Windscale. Discreet samples were also taken around the Lancashire coast, the North Wales coast, the Isle of Man and into Yorkshire and the south of Scotland. As the evidence came in, the authorities had to revise their threat estimates. The restriction on distribution was extended in successive stages, until by Monday morning it covered a wide area from a coastal strip approximately fifty kilometres long, sixteen kilometres broad at the southern end and ten kilometres broad in the north. In the south it included the Barrow Peninsula with the northern boundary about ten kilometres north of Windscale.

As the news spread, public anxiety grew. The Medical Research Council Committee issued a soothing report stating that, 'It is in the highest degree unlikely that any harm has been done to the health of anybody, whether a worker in the Windscale plant or a member of the general public.' An emollient statement and simply not true. The Windscale fallout actually caused at least 240 cancers in Britain, half of them fatal.

The great clear-up – and the great Whitehall cover-up – began quickly. Following the fire, the British government downplayed the severity of the accident to the press. On 15 October, it was announced by the chairman of the UKAEA that a Committee of Inquiry had been established under the chairmanship of Sir William Penney to conduct an investigation into the accident. The Committee sat at Windscale during 17–25 October, interviewed thirty-seven people (some more than once), and examined seventy-three technical exhibits. There were interviews with key scientists and operators from the time. Previously undisclosed material was used, including taped interviews conducted directly after the fire.

The main focus of the enquiry was technical, scrutinising the workings of the reactor pile and who did what, and when. The environmental consequences and the pressure from the British government were hardly mentioned, although the chairman is recorded as exchanging a jokey comment about the difference between cows' milk and lactating women's breast milk in the local area, including the gem, 'But then, mothers don't eat grass, do they?'

Penney's report was handed to the chairman of the UKAEA on 26 October. It went into considerable detail on the incident, blaming the staff rather than the plant's structures and the operations themselves. The principal conclusions were:

- The cause was human error.
- Technicians mistakenly overheated Windscale Pile No. 1.
- Poorly placed temperature sensors indicated the reactor was cooling rather than heating.
- The excess heat led to the failure of a nuclear cartridge, which in turn allowed uranium and irradiated graphite to react with air.
- The resulting fire burned for days, damaging a significant portion of the reactor core.
- About 150 burning fuel cells could not be lifted from the core, but operators succeeded in creating a firebreak by removing nearby fuel cells.
- An effort to cool the graphite core with water eventually quenched the fire.
- The reactor had released some radioactive gases into the surrounding countryside, primarily in the form of iodine-131.

There was no mention of the British government's rush to create extra materials for a British nuclear bomb, the order to trim the cooling fins from the rods or the other safety corners that had been cut. Although it was effectively a whitewash, the report still caused alarm in Whitehall. Prime Minister Macmillan told a Cabinet meeting that, after carefully studying the report, he had come to the conclusion that its publication would not be in the public interest.

A heavily censored version of the 'Penney Report' formed the technical basis of a UK Government White Paper published on 8 November 1957, but the full and damning account itself was never published, and only made public in 1988 under the Thirty Years Rule. The released documents show that not only was the original report suppressed, but major efforts were made to ensure all circulating copies were returned. According to the official classified documents, Prime Minister Harold Macmillan had ordered the report on Windscale to be rewritten after AEA members said they were concerned about its findings.

The 1988 disclosures brought strong criticism from the opposition Labour Party. 'The main lesson of all this,' said the late Tony Benn, a former Energy Secretary, 'is that in the field of atomic power and nuclear weapons the British people have never been told the truth.'

Sixty years on at Windscale, the mud beaches down towards the Irish Sea still test slightly positive for residual radiation. The original primitive Windscale reactor has been carefully dismantled and the site of Britain's first and worst nuclear accident has been incorporated, along with Calder Hall, into the multi-purpose nuclear facility Sellafield.

Whitehall has never provided a trustworthy official record of the true number of Windscale casualties.

CHAPTER 16

THE WORLD HELD ITS BREATH: CUBA 1962

'Within the past week, unmistakable evidence has established the fact that a series of offensive missile sites is now in preparation on that imprisoned island. The purpose of these bases can be none other than to provide a nuclear strike capability against the Western Hemisphere.'
– PRESIDENT JOHN F KENNEDY

If the Berlin Crisis – the major political stand-off between the Eastern and Western Blocs occupying the two parts of Berlin – in 1961, and the erection of its infamous dividing wall that same year, were not bad enough, no other single event brought home the real dangers of the nuclear arms race and the Cold War as the Cuban Missile Crisis. It's a well-worn phrase, but the world did indeed hold its breath in October 1962.

To those who lived through it, who have memories of preparing to go to nuclear shelters, reading about the preparation of nuclear weapons for war and wondering if it would actually happen, the Cuban Missile Crisis was a real frightener. Millions of lives all over the world teetered on the edge of a precipice. 'Brinkmanship' is another well-used buzzword often applied to the US–USSR stand-off, a strategy deployed by both sides as they sent their nuclear weapons into the front line. Two men held the fate of the world in their hands.

The crisis had its origins in two fundamental factors: the need for Nikita Khrushchev to demonstrate his firmness to his fellow Communists against the 'principal adversary'; and the misunderstood relationship between the Soviet leader and the youthful President John F Kennedy. And all this against the backdrop of the tensions over Berlin and inflammatory rhetoric on both sides.

For three years, Khrushchev had been sabre-rattling unsuccessfully over Berlin to try to achieve some solution to what the Kremlin saw as the problem of an increasingly powerful West Germany. From the Soviet perspective, the whole point of the victory in the Great Patriotic War had been to keep Germany weak and subservient. Krushchev was also under pressure at home from political hard-liners. By the early 1960s, the Sino–Soviet split had widened. Despite massive H-bomb tests that demonstrated the power of the USSR, he was keenly aware of growing US technological superiority.

Everywhere Khrushchev looked, the Soviet Union was being encircled by the USA or its NATO allies, threatened from across the polar ice cap and the Bering Straits in the north, in Berlin and from the Black Sea to Bokhara (now Bukhara) in the deep south. Psychologically, Khrushchev and the USSR felt surrounded and ensnared.

The volatile and impulsive Ukrainian appears to have decided that he had to act to show his strength. What better way to demonstrate both his commitment to revolutionary Communism and the USSR's power than by supporting Fidel Castro's Marxist-inspired socialist regime in Cuba? And at the time, the Cuban revolution was in jeopardy from US Marines, now massing in the Caribbean preparing for an invasion. Soviet intelligence had, quite accurately, alerted him about top-secret

US plans emanating from the White House itself, to mount a coup and topple the Cuban regime. Revolutionary sentiment, rivalry with the USA and a belief that the new US president was a shallow, weak and ineffectual adversary, lured the Soviet leader into his most dangerous gamble of all.

Khrushchev felt that he could take advantage of the boyish and inexperienced president. He had forced Kennedy into a humiliating climbdown over Berlin and had openly mocked the disastrous US-backed Bay of Pigs debacle in Cuba in April 1961, in which a CIA-led invasion force to topple Castro was spectacularly defeated within twenty-four hours.

Although the circumstances of his tragic death have given us a dewy-eyed memory of Kennedy and his much-hyped 'Camelot' (the term his widow, Jackie, coined to describe her husband's presidency), the hard fact remains that, in 1962, Khrushchev regarded him as a lightweight. The stage was set for an apocalyptic, misjudged confrontation.

It now seems clear that the CIA warned Kennedy in vague terms of some Soviet adventure brooding in Cuba. Surely, even Khrushchev wouldn't be so stupid as to provoke the USA by sending strategic weapons, let alone offensive nuclear missiles?

Khrushchev knew that the US Jupiter missiles in Turkey could reach Moscow from the very border of the USSR. The Soviet premier decided, to translate from a vivid Russian phrase, 'to put a hedgehog down Kennedy's pants'. He would place Soviet nuclear missiles just across the US border and see how they liked that. Strategically it made sense too. Nuclear missiles close to continental USA were unstoppable and would discourage any attempts at a nuclear first strike by the notoriously hawkish US Air Force generals.

The plan was for 50,000 personnel and their heavy weapons to deploy to Cuba by sea. Soviet diplomats prepared a cover story by boasting of a major civilian development programme in Cuba. In late August, a US spy flight spotted new construction near Havana, but analysts correctly identified the sites as SAM-2 anti-aircraft missile pads. The President assured Congress that there were no offensive missiles on Cuba.

During the night of 8 September, forty SS-4 and SS-5 nuclear Medium Range Ballistic Missiles (MRBMs) were unloaded at Havana. SAM-2 anti-aircraft missiles would protect the launchers, along with the normal complement of rocket support troops. This coincided with one of those inexplicable gaps in the United States' usually watchful eye on potential sources of trouble. For over a month it seemed no one bothered to monitor events in Cuba. Five weeks later, however, American intelligence made an alarming discovery.

On 14 October 1962, a U-2 intelligence collection flight brought back film of new concrete missile launch pads at San Cristóbal in western Cuba. The startled intelligence analysts at Washington's National Photographic Interpretation Center (NPIC) and the Pentagon couldn't believe their eyes. The Soviets were setting up a nuclear missile base right under their noses.

On the morning of 16 October 1962, the President was informed that the USSR was deploying nuclear missiles 145 kilometres (90 miles) from the Florida coast. 'My God,' said a shocked Kennedy. 'Why that's as bad as if we'd suddenly begun to put a major number of missiles into Turkey!' An aide reminded him, 'Well, if you remember, that's just what we did do, Mr President...'

The Cuban Missile Crisis had begun.

Although Castro had originally been reluctant to have Soviet nuclear weapons based on Cuban soil, once he had accepted Khrushchev's idea he became an enthusiastic supporter. The USA really *was* limbering up to invade Cuba in 1962. The humiliated Kennedy White House was boiling with rage over the Bay of Pigs fiasco and was conspiring with Mafia leaders and the CIA to get the island back for the Cuban émigrés and the East Coast Mob. Declassified top-secret documentation shows only too clearly just how determined was the USA, and especially the Pentagon hawks, to eliminate this irritating gadfly just off Florida.

During the presidential race of 1960, one of Kennedy's key election issues was an alleged 'missile gap', an area in which the USA supposedly trailed the Soviet Union. In fact, the Americans were ahead by a considerable margin, a margin that would widen still further; while the Soviets were in possession of just four intercontinental ballistic missiles in 1961 (the R-7 Semyorka), increasing to perhaps a few dozen by the following year, the USA already had 170 ICBMs, with more to follow. It also boasted eight ballistic missile submarines, which were the equal of the *George Washington* and *Ethan Allen*, each one capable of launching sixteen Polaris missiles, which had a range of some 4,600 kilometres. Khrushchev may have publicly announced that the Soviets were constructing missiles 'like sausages', but the true numbers were far more modest.

Added to the wave of 'piracy' off Cuba's coasts (attacks by die-hard anti-Castro Cuban émigrés), it is clear that Castro, and Khrushchev, had every reason to fear some US-led intervention. The militarists in the Pentagon were undoubtedly readying themselves for action against Cuba when the Soviet Union offered to 'protect the Cuban revolution from imperialist aggression'. The Americans' discovery that this now meant something much more dangerous than a handful of clapped-out IL-28 Beagle bombers and some fast patrol boats, raised the stakes dramatically.

US Under Secretary of State George Ball even described the discovery of the Soviets' nuclear strategic missiles on Cuba as a psychological shock comparable to that of the attack on Pearl Harbor in 1941. President Kennedy's first act was to convene a small 'war cabinet'. Known as 'ExComm' (for the Executive Committee of the National Security Council), this group effectively managed the situation over the next two weeks. At the time, such crisis management was an innovation. Even more so, to many of the participants, was Kennedy's secret taping of the meetings. Few of those present at the time realised that their words and deeds would become one of the most studied pieces of decision-making in history.

The 'shocked incredulity', in the words of Robert F 'Bobbie' Kennedy (US Attorney General and the President's younger brother), that the Soviets would have the cheek to risk such a venture soon dissipated with the knowledge that, according to the Pentagon, the Soviet missiles would be active and ready for firing in just fourteen days. ExComm's options crystallised rapidly: do nothing (swiftly rejected); a blockade; invasion; a 'surgical' air strike (much favoured by the US Air Force generals, desperate to flex their muscles); direct talks with Castro; and diplomatic negotiations with Moscow.

In the end, only two options were left open, apart from the obvious need to start talking to the Soviets: a blockade or a combined air strike and invasion. All the options had serious drawbacks. An air strike and invasion risked not only Soviet retaliation and another crisis elsewhere (for example, Berlin), but it smacked of 'Pearl Harbor in reverse' and sat uneasily with American ideas of morality and surprise attacks.

Kennedy was also woefully misinformed about the size of the Soviet troop presence on Cuba. On 20 October, following the discovery of the missiles at San Cristóbal, US Defense Secretary Robert McNamara told the president that there were 6,000 to 8,000 Soviet 'technicians' on the island. In fact, there were no less than 43,000 heavily armed Soviet troops on Cuba, equipped with tactical nuclear weapons targeted at suspected US beachheads. Kennedy rightly rejected as too risky calls by the Joint Chiefs of Staff (JCS) to invade, even though he didn't know the half of it.

The JCS recommended a massive US air attack on Cuba, to be followed by an invasion within seven days, which we now know could have resulted in tens of thousands of US, Soviet and Cuban casualties, the nuking of the Guantánamo Bay Naval Base and, quite possibly, full-scale nuclear war. We can only be grateful for JFK's restraint.

In the middle of all this frenzied debate behind the scenes, the President had a long-scheduled routine meeting with Andrei Gromyko, the Soviet Foreign Minister. The meeting was diplomatically cordial, if guarded. Neither side raised the issue of the Soviet missiles, although Kennedy actually had copies of the secret aerial photographs in his desk drawer should he need to challenge the Russian. Eventually, Gromyko accused the USA of 'harassing a small country – Cuba'. In reply, Kennedy assured the Foreign Minister that the US had no designs on Cuba but he had noted that the Soviets were encouraging a military build-up on the island. Gromyko countered by agreeing, but insisted that Soviet arms shipments were purely defensive. No Soviet offensive weapons had been, or would be, sent to Cuba.

Both men were lying in the finest traditions of international diplomatic relations. In his turn, Gromyko went away delighted and reported to the Kremlin that the US appeared to have no inkling of the Kremlin's plans. 'A US military adventure against Cuba is impossible,' he reported.

In a week or so, Soviet Cuban missiles would be ready. Kennedy was facing mid-term elections in November. The Russian leader would soon be able to spring a political and diplomatic triumph that would seal his place in history as the undoubted leader of the Communist world, and the man who finally checkmated the Americans in their own back yard.

But Kennedy had now made up his mind: he was going to blockade Cuba. This course would give the Russians time to think things through. The problem was that a blockade is either illegal or an act of war. What if the Soviets ordered their ships to try and run the US Navy blockade? Would the navy have to use armed force to stop them? On 21 October, Kennedy finally authorised the action to impose a 'quarantine zone' around Cuba, observing gloomily, 'There isn't any good solution . . . but this one seems the least objectionable . . .' The next day, the President broke into prime-time television schedules to tell the American people the grim news.

Behind the scenes, an embarrassed Anatoly Dobrynin – the Soviet Ambassador to the USA – was being carpeted by US Secretary of State Dean Rusk at the State Department. Rusk handed him a copy of Kennedy's public announcement and a personal letter – amounting to an ultimatum – to Khrushchev. The unfortunate Dobrynin nearly collapsed with shock. By the time he managed to signal Moscow, the Cuban Missile Crisis had erupted as a major threat to world peace and US nuclear forces had gone on to a war alert as a precaution.

No one seems to have been more surprised than Khrushchev himself. The whole Soviet leadership seems to have been caught on the hop. 'This could end in a very big war . . .', Khrushchev advised a startled Presidium, the USSR's top-ranking executive committee. The Party comrades were taken aback. Khrushchev had led them to agree to the whole Cuba adventure by assuring them that there was no risk. The Kremlin now tried to work out what to do next.

Having been caught out, Khrushchev immediately turned to his favourite tactic – bluster. He alerted the Warsaw Pact, called Kennedy a warmonger and accused the USA of piracy on the high seas. Yet Khrushchev was as scared as anyone about the consequences of his reckless scheme, now that it had been uncovered. War hero General Vasily Kuznetsov observed dryly that, 'Khrushchev shit his pants' with fright.

Meanwhile, twenty-six Soviet merchant ships ploughed across the Atlantic towards Cuba. On 24 October, the US picket line of warships were on station, ready to bar their way. A confrontation was inevitable. Khrushchev sent a message that if any Russian ships were stopped, then Soviet submarines would attack the US warships to protect their own shipping. 'If the US insists on war, then we'll all meet in hell.'

The first confrontation came on 25 October, when a destroyer ordered the Soviet tanker *Bucharest* to heave to. But the *Bucharest* was only carrying oil and after inspection was allowed to proceed to Havana. On the bigger international stage, things had gone from bad to much worse. Khrushchev was now accusing the USA of open aggression and had put Soviet nuclear forces on alert. In response, the US Air Force Strategic Air Command had gone to 'DEFCON 2', the US military's next-to-hostilities state of readiness for war, with twenty-four-hour cab ranks of nuclear-armed B-52 bombers orbiting their fail-safe points over the polar ice cap. American ICBM missile silos were prepared for firing. Nuclear submarines sailed to their war stations and made ready to launch their nuclear missiles. In Cuba, the Soviet commander reported that his SS-4 missiles would be ready for launch within a day. The US Marines prepared to invade Cuba once the order was given.

The same day saw the diplomatic nadir of the entire episode. In a UN Security Council debate, the Soviet delegate denied that the USSR had any missiles in Cuba and demanded evidence of the US claims. In a dramatic *coup de théâtre,* Adlai Stevenson II, the US delegate, revealed the top-secret aerial photographs of the SS-4 installations. The Soviets had been caught lying, and in the world's most public forum. The embarrassed Soviet delegation denied the American claims, saying that the photographs were 'obvious forgeries'. It was against this uncompromising background that Kennedy and his ExComm now reviewed the first of two letters from Khrushchev.

Aleksandr Fomin, a KGB officer (real name Aleksandr Feklisov) masquerading as a journalist, called up John Scali, the diplomatic correspondent for ABC TV, and offered a compromise deal between the USA and the USSR. If America promised not to invade Cuba, then Moscow would withdraw its missiles: could Kennedy deal?

Khrushchev, who by now appears to have calmed down, had been authorised by the Presidium to negotiate and seek a diplomatic solution. A contemptuous Kremlin brushed aside an encouraging suggestion by Castro that the Soviets should go for broke and consider a nuclear first strike. Peace, not a nuclear war, was Moscow's preferred outcome. Events, however, conspired against the doves. A long-scheduled Atlas missile test launch alarmed the Soviets. An unauthorised high-altitude U-2 probe over Siberia had drawn swift retaliation from the nervous Russian air defences and the pilot had been lucky to escape. The Soviets took it as a prelude to war.

The next day, 27 October, brought even worse news. A USAF U-2 had been shot down over Cuba by a Soviet SAM-2 and the pilot was dead. As their contingency plans dictated, the US military now prepared to strike back at Cuba, which had effectively opened fire on them first. As ExComm glumly contemplated the nuclear stand-off – with Cuba on the brink and a confrontation at sea off Havana in the making, a U-2 shot down over Cuba and war seemingly inevitable – they received a second message from Khrushchev, much tougher than the first.

We now know that it was an American climbdown that finally broke the Cuban stand-off. The confirmation comes from a telegram Ambassador Dobrynin sent to the Soviet Foreign Ministry, Khrushchev's memoirs and lawyer Theodore Sorensen. The latter was also the President's speechwriter, close adviser and the uncredited author of *Thirteen Days*, which presented the White House version of the Cuban crisis. (It was published under Robert Kennedy's name.) The night before Khrushchev's second message was received, a worried Robert Kennedy had met Ambassador Dobrynin in secret. From the KGB Archive we now have access to Dobrynin's own top-secret telegram to the Kremlin, dated 27 October. His account of the meeting is confirmed by both Khrushchev's memoirs and by the 1989 'confession' of Sorensen at a special conference in Moscow attended by those involved on both sides.

Dobrynin's telegram gives an unchallenged version of events that conflicts with 'JFK's Great Triumph'. According to an account by Dobrynin that was preserved in the archives of Russia's Foreign Ministry, Robert Kennedy panicked. He came to him 'almost in tears', admitting:

> Because of the plane ... that was shot down, there is now strong pressure on the President to give an order to respond with fire if fired upon ... The USA can't stop these flights, because this is the only way we can quickly get information about the state of construction of the [Russian] missile bases in Cuba ... [Kennedy added that there were many US generals, and others high up in government, 'itching for a ... fight'.]
>
> [The USA sought] the agreement of the Soviet government to halt further work on the construction of the missile bases ... in Cuba ... In exchange the government of the USA is ready, in addition to repealing all measures on the 'quarantine' [blockade of Cuba], to give the assurances that there will not be any invasion of Cuba ...

'What about Turkey?' Dobrynin asked RFK, in a key exchange that Sorensen later admitted that he had deliberately omitted from *Thirteen Days*. 'If that is the only obstacle ... the President doesn't see any unsurmountable difficulties in resolving this issue ...' came the reply. 'The greatest difficulty for the President is the public discussion of the issue of Turkey ... To announce now a unilateral decision by the President of the USA to withdraw missile bases from Turkey would ... damage the entire structure of NATO and the US position as the leader ... of NATO ...

I think that in order to withdraw these bases from Turkey we need four to five months . . . However, the President can't say anything public in this regard about Turkey.'

RFK had quite explicitly offered to trade the US position in Turkey as a quid pro quo for the SS-4s being taken out of Cuba. The Americans wanted out of the crisis and were prepared to make concessions. Dobrynin cabled Moscow. A relieved Kremlin accepted.

The problem was that neither the US military nor ExComm were aware of this dangerous secret backstairs bargaining by the President and his energetic young brother. When Khrushchev's second letter arrived insisting 'that the Turkish missiles must be removed' as part of a deal, the generals and warmongers now believed that war was inevitable and argued for a US first strike.

The Kennedy brothers wriggled and dodged uneasily in the tense discussion that followed. In the end, ExComm (slightly to its surprise, from the transcripts) decided that the best course might be to pretend that they had not yet seen the second letter and replied to Khrushchev's first, milder offer, although the hawks around Kennedy still advocated war.

It was Kennedy's reply to Khrushchev that broke the crisis. In it he promised to call off the US blockade if Russia removed its missiles from Cuba. However, and crucially, behind the scenes the US President and his brother secretly promised to remove the US Jupiter missiles from Turkey as well. That was the real deal.

If it came out however, the political fallout with the American voters and the bellicose military, would damage the Kennedy presidency beyond repair. It would look as if *they* – and not Khrushchev – were the ones who had 'blinked'. That would never do.

Therefore, to cover their tracks, and to save face should it ever come out, the Kennedy brothers secretly primed U Thant – the

UN Secretary General – with a cover story. They asked him to be ready to claim that it was *he* who had suggested this scheme, should the Soviets ever let the cat out of the bag to try to grab any propaganda advantage from the real deal, which was a secretly agreed mutual climbdown. The White House must not be seen to have conceded ground.

A delighted Khrushchev agreed and on 28 October ordered his ships to carry the offending missiles back to the USSR. The missile sites would be dismantled and UN inspectors would confirm that the Russians on Cuba had gone home. Khrushchev had succeeded in getting rid of US missiles threatening Moscow.

The Americans crowed in public. A grateful world agreed and heaved a collective sigh of relief. Kennedy, the boy president, had stared down Khrushchev and saved the world. Or that was the story. Not everyone was fooled. Khrushchev pointed out with some justice to his shaken colleagues on the Presidium that he had achieved his aim: a guarantee that the imperialist Yankees would not now invade Cuba. And the Americans had now promised to dismantle their missiles in Turkey (and, away from the glare of publicity, the Jupiters in the UK and Italy as well) as a bonus. It was a Soviet diplomatic and Cold War triumph, crowed the Russian leader. His brilliant plan had worked.

Many inside the American military tended to agree. They were unimpressed with Kennedy's great diplomatic triumph. Insiders were far from convinced that 'the other guy blinked'. 'We've been had . . .', observed one laconic US admiral. And he wasn't just referring to the Russians.

Back in Moscow the ever-paranoiac Presidium of the Communist Party of the Soviet Union eyed their impulsive and crude leader with increasing distaste. His lunatic – and completely unnecessary – scheme to embarrass the Americans had nearly killed them all.

At the time, everyone was so glad to survive that they failed to examine Kennedy's 'statesman-like triumph' too closely. The reality of the Cuban Missile Crisis of autumn 1962 was that – somehow – both sides had managed to prevent a disastrous nuclear exchange. The true victory was that both Washington and Moscow had achieved their goals without going to war.

Peace was the real winner.

CHAPTER 17

THE REAL CUBAN MISSILE CRISIS – AT SEA

'For the last four days, they didn't even let us come up to the periscope depth . . . My head is bursting from the stuffy air . . . Today three sailors fainted from overheating again . . . the carbon dioxide content [is] rising, and the electric power reserves are dropping. Those who are free from their shifts, are sitting immobile, staring at one spot . . . Temperature in the sections is above 50°C.'
– **ANATOLY ANDREEV, SOVIET SUBMARINER OFF CUBA, 1962**

Most people believe that the world never came closer to nuclear Armageddon than during the Cuban Missile Crisis in October 1962. In popular memory, the decision for war would have come from Kennedy and Khrushchev sitting in Washington or Moscow, as the United States and Soviet Union confronted one another over the USSR's deployment of ballistic nuclear missiles to Cuba.

The truth in 1962 was that the real danger came not from any intercontinental exchange of ICBMs. The decisions that prevented World War III were actually being taken by some very frightened men in the Caribbean, 2,000 miles (3,200 kilometres) to the south of leafy Washington. At the heart of the action was a dangerous maritime confrontation between US destroyers and Soviet submarines.

Previously, Moscow had sent a convoy to Cuba, one which comprised eighty-six Soviet ships with a motorised rifle division. The Soviets had also despatched forty MiG-21 jet fighters, sixteen ballistic launchers of R-12 and R-14 missiles, a dozen FROG-3 ballistic missile systems that were nuclear capable, half-a-dozen Il-28 jet bombers and two anti-aircraft divisions with SA-2 surface-to-air missiles (SAMs). Troops and equipment had been hidden from view on the ships, though there was one occasion, on 4 September 1962, when the US Navy aircraft were able to identify some of the SAMs on one transport. On the whole, however, the Soviet *maskirovka* ('deception') operation was a remarkable success.

It was far more difficult, however, to deploy, undetected, such a large force on land, and on 14 October a US U-2 spy plane photographed the Soviet ballistic missiles at San Cristóbal. This triggered the missile crisis, and Kennedy ordered a naval blockade of Cuba one week later, an operation involving hundreds of ships, among them four aircraft carriers, plus many other patrol planes on the shores. While the Soviets appeared to defy the blockade, a relatively small number of their ships persisted, with most of their ships performing U-turns. Yet weeks earlier, they had activated 'Operation Kama', during which they had escorted the transport ships to Cuba by deploying four Foxtrot-level diesel submarines from the 69th Torpedo Submarine Brigade. Overseeing these submarines, which were falling into increasing states of disrepair, was Chief of Staff Vasili Arkhipov. A year earlier, Arkhipov had been executive officer of the *K-19* and had only just averted a nuclear meltdown off Greenland. In the ensuing chaos, he had been severely irradiated, but – unlike many *K-19* crewmembers – survived until 1998.

Between them, the Soviet submarines and the US Navy hunter-killer anti-submarine groups were unwittingly to act out the real nuclear incident waiting to happen off Cuba. Theirs was a potentially deadly naval game of blind man's buff, played out beneath the waves off Cuba's eastern approaches. The men who would take the key decisions were not sitting in cool, dark rooms discussing the latest developments. Instead, the decisions were being taken by a group of exhausted, frightened men in the throes of dehydration and CO_2 poisoning as they sat in a malfunctioning submarine, unable to consult with Moscow and surrounded by nervous US destroyers. At the heart of the confrontation was a potentially lethal intelligence failure. Both sides were fielding tactical nuclear weapons and were preparing to use them. And neither side knew it.

When Defense Secretary McNamara became aware – on 23 October – that the Soviet freighters were being escorted by the Soviet submarines, he authorised US ships to warn them off with 'practice depth charges' (PDCs). The size of grenades, PDCs had been designed to be used in training, or as a way of informing the submarines that they were being detected. They may have been small, but their impact was great enough to damage the radio antennae of the Soviet submarines, whose crews believed they were under full attack. While the USA had informed Moscow of its 'Submarine Surfacing and Identification Procedures', clearly the details never reached the 69th Brigade.

The US Navy knew that they had to interdict the Soviet boats below the waves. The question was, how? They couldn't attack them: that would have meant war. The navy realised that Soviet submarines were trying to run the blockade ahead of their merchant ships and had to contact and stop them. However, while it is relatively straightforward to stop a surface vessel and order it to heave to, submarines are a very different matter. The US Navy was now a major player in the dangerous game of maritime deterrence. Under direct orders from the Pentagon, US destroyers, submarines and aircraft began to track the Soviets' diesel-powered submarines and attempted to signal them to stop.

As the Americans closed in on their quarry, far below them the frightened Russian submariners began to load nuclear-tipped torpedoes. The Americans hadn't a clue that these weapons even existed. On the surface, US anti-submarine carriers loaded their nuclear depth charges, just in case. This was important, because the Soviets didn't know about *those* either.

Valentin Savitsky, submarine *B-59*'s commanding officer, could not have known that the USS *Beale*'s depth charges were merely warning shots designed to force his submarine into surfacing. Other US destroyers joined in the mini-attack on the *B-59* with further depth charges and live sonar signal blasts. Finally, Savitsky, under the impression that World War III was fully underway, ordered the *B-59*'s 10-kiloton nuclear torpedo to be unleashed, with the USS *Randolph* aircraft carrier as its intended target.

Had the *Randolph* been attacked, the nuclear clouds would have spread to the land, and would probably have triggered the Pentagon's Single Integrated Operational Plan (SIOP). This would have resulted in the kind of situation satirised in Dr Strangelove's fictional *Götterdämmerung*: nuclear weapons firing towards targets in the Soviet Union, China and many other places besides. In retaliation, Soviets could have destroyed many British targets – London and air bases in East Anglia, for example – and concentrations of troops in Germany. A further bombing campaign would have wiped out 'economic targets' (meaning civilians), resulting in the deaths of over half the population of Britain.

Known as Project 641s to the Soviets, the Foxtrot submarines had been introduced back in 1957. With their loud and rattling propellers, Foxtrots lacked the teardrop-shaped hulls of later models, which were stronger and faster, and their third decks contained vast batteries, which enabled them to operate below the surface of the water for ten days. All this meant that a seventy-eight-person crew had minimal living space, and the submarine's cooling systems were not really suitable for tropical waters.

On 1 October 1962, the Foxtrots set out from the Kola Peninsula, evading the NATO's anti-submarine aircraft and reaching the North Atlantic, but towards Cuba, they needed to surface in order to recharge the crafts' batteries, and in these more tropical waters, living conditions underwater became more and more difficult. When the cooling systems faltered, temperatures rose to between 38 and 60°C, and a build-up of carbon dioxide began to affect the health of everyone on board, as did the lack of fresh water, which led to dehydration and rashes.

Above them, the US Navy was surrounding their quarry, dropping their PDCs.

Hearing the explosions, pandemonium broke out in the Soviet submarines. They had been at sea for weeks; the crews were tired, frightened and very jumpy. After being detected, Savitsky and his exhausted crew bravely endured four more hours of US active sonar blasting sound waves against its hull. The sub's frightened crewmen were dropping like flies. Their air-conditioning equipment was inoperable. Temperatures soared and the air was turning foul.

The clang of approaching American depth charges was the final straw. On at least two of the Soviet boats, the captains prepared to fire nuclear torpedoes at the American ships. On submarine *B-130*, the skipper actually ordered the nuclear tube to be loaded, then watched with wry satisfaction as his *zampolit* ('political officer') wet himself and fainted with fright. Captain Nikolai Shumkov grunted and ordered the nuclear torpedo (which was fifteen times the size of the Hiroshima bomb) to be unloaded. 'Just testing,' he told the stunned control room of *B-130*, adding that he had no intention of using the torpedo, 'because we would all go up with it if we fired. Agreed?' His wide-eyed officers dumbly nodded their heads in agreement.

The tension mounted in submarine *B-59* as well. Witnesses claim that Captain Savitsky shouted, 'Maybe the war has already started up there. We're going to blast them now! We'll die, but we will sink them all. We will not disgrace our navy.'

The *B-59*, however, had three senior officers on board. And Captain Savitsky was not the most senior among them – political officer Maslennikov and Acting Commodore Vasili Alexandrovich Arkhipov could issue orders as well. The man who saved the day on 27 October 1962 was the senior Russian submarine officer, Arkhipov.

The key point was that they were only authorised to fire the 'special weapon' if all three unanimously agreed to do so. As Arkhipov was the flotilla commander, *B-59*'s captain and political officer needed to get his approval. Alone, Arkhipov coolly opposed the launch and worked to persuade Savitsky to surface the submarine to find out from Moscow exactly what was going on in the wider picture.

As the submarine's batteries had run very low and its air-conditioning had failed, *B-59* finally had no option but to surface in order to use its diesel engine. She did so, to find herself surrounded by the pursuing US warships, which provided a surprising welcome. US Navy Ensign Gary Slaughter of the destroyer USS *Cony* later recalled the events that followed when the submarine emerged from the deep at 9:00 pm that night to be bathed in light blazing from the American vessels. The crewmen poured out of the hatches, overwhelming relief and joy evident on their faces. Slaughter was the only *Cony* officer who could communicate with the Russian captain, and did so with the help of the Cyrillic Transliteration Table, Morse

Code and the International Signals Book. What followed was a tense stand-off between Slaughter and Savitsky, who would only name his ship as 'Soviet ship X', and his craft's status as 'On the surface, operating normally'.

After an hour, a navy pilot in a P2V Neptune broke the silence by dropping several incendiary devices in order to activate his photoelectric camera lenses. Savitsky, in a panic, turned his craft around, torpedo tubes at the ready, but after an apology from Slaughter, returned to port. Slaughter was ordered to keep relations as cordial as possible with Savitsky, and relations improved to the point where Savitsky signalled his crew would welcome fresh bread and US cigarettes. Slaughter complied with his request, the package was received with thanks and the *B-59* began its journey back to the Soviet Union.

'All crew members were sworn to secrecy about the entire *B-59* incident,' revealed Slaughter in his subsequent account. 'While we didn't know why, we took our vows of silence seriously. For the next forty years, I never told anyone about this incident.'

Only years later did it occur to Gary Slaughter that what he had witnessed off Cuba that night aboard the USS *Cony* was not only the culmination of the Cuban Missile Crisis but the day that really saved the world.

Nuclear war could easily have broken out during the Cuban Missile Crisis, not as part of some strategic exchange of ICBMs, but at sea and almost by accident, merely as a very low-level tactical exchange of nuclear weapons between surface ships and submarines crewed by exhausted and frightened men. That is the really disturbing aspect of those terrifying days in the autumn of 1962.

Nearly sixty years on, what is to be learned from the Cuban Missile Crisis? Crucially, we must realise that governments can lose control in such circumstances. Robert McNamara faced the possibility of an unauthorised launch of a nuclear weapon, and his reaction was to decree that Permissive Action Links (PAL) locks must be fitted to all ICBMs. But when they were installed, Curtis le May – the devious boss of USAF Strategic Air Command – arranged for all the SAC codes to be set to 00000000, in order that the official locks would not obstruct a quick launch if deemed necessary.

As with all such matters, security concerning nuclear weapons will forever be a human issue. Even Jimmy Carter, so often regarded as one of the most thoughtful and careful of US presidents, once sent his suit to the dry cleaners with the nuclear launch codes accidentally left inside.

The second lesson is that events on the ground soon overtake political aims. A routine test launch here, or an accidental incursion there, send the wrong signals and are capable of the worst-case interpretation by nervous politicians. Unknown to most, the world came very close to a nuclear war in 1962. A Soviet naval commodore probably saved the world.

CHAPTER 18

EXERCISE ABLE ARCHER 83

'Ended? But it's only day 3 of the exercise. What the hell is going on? I bet those idiot politicians have ****** up again. Oh well, at least we get home early for once . . .'
– **OBERST VON TZSCHIRNER, BUNDESWEHR GENERAL STAFF, SHAPE INTELLIGENCE DIVISION**

There is general agreement – and plenty of evidence – that the nearest the world came to a full-blown accidental strategic nuclear war was during 'Exercise Able Archer' in late 1983. The 2015 release of the full US classified papers proves this without doubt. Thanks to serious misunderstandings by both sides, a full nuclear war was a very real danger.

Towards the end of the 1960s, NATO began organising a series of military exercises in Europe that aimed to train its forces and assess standards of readiness for combat. The tests, dubbed 'Autumn Forge' as they took place between each September and November, were run by NATO's Supreme Headquarters Allied Powers Europe (SHAPE), and by 1983 – the year of the Able Archer war scare, around 100,000 troops were participating in half a dozen exercises. The largest of these was Reforger 83 (the name deriving from 'REturn of FORces to GERmany'), which required NATO forces to defend against the 'Orange Pact' (or Warsaw Pact), and which airlifted over 16,000 troops from the USA to Europe.

The last Autumn Forge exercise, 'Able Archer', assessed both command and control procedures with regard to conventional and nuclear weapons. 'It was an annual Command Post Exercise (CPX) thus involving only headquarters, not troops on the ground, of NATO's Allied Command Europe (ACE),' recalled SHAPE historian Gregory Pedlow, 'and it was designed to practise command and staff procedures, with particular emphasis on the transition from conventional to non-conventional operations, including the simulated release of nuclear weapons.' It was just a communications exercise.

Quite how the USSR was panicked this time remains a mystery. Neither Ambassador Dobrynin, Oleg Gordievsky – Britain's double agent in the KGB – nor the head of the KGB residence in Washington at the time, Oleg Shvets, can throw any light on the matter. A simple explanation might be that the Soviet Premier at the time, Yuri Andropov, was a former head of the KGB. A certain *déformation professionelle* is to be expected and becomes almost inevitable with long-time intelligence officers, whose view of life, like policemen, sometimes turns out to be warped by their experience. For many, suspicion, even paranoia, becomes their normal view of life.

The Americans deliberately fed these Soviet insecurities. Washington launched enough surprises to worry anyone, let alone the nervous old men in the Kremlin. When the Reagan administration came to power, one of their carefully planned offensives against the USSR was a deliberate – and still highly secret – campaign of psychological operations (PSYOPS) designed to deliberately jangle Soviet nerves. Premiers Andropov and Chernenko both seem to fallen victim to their fears, and their nerves were as easily jangled as any KGB analyst. The US PSYOPS campaign of the 1980s was, therefore, a success, but had some unintended consequences. Few people know of this secret campaign to destabilise and worry the Soviet leadership, and the details have still never been fully revealed.

The programme was intended not to demonstrate American intentions but to keep the Russians guessing as to what might come next. Bombers would swoop over the North Pole and turn away when alarmed Soviet air defence radars were just about to switch from 'detection' to 'target illumination' mode. The US Navy also stepped up its activities as part of the PSYOPS campaign, beginning with a massive ninety-ship NATO Fleet exercise off the Soviet Red Banner Fleet's headquarters in Murmansk in the autumn of 1981.

During the next twelve months, US carrier jets practised aggressive air-to-air interceptions of Soviet maritime patrol aircraft and demonstrated that they could evade Soviet warning radars by popping up in unexpected places. US and British nuclear submarines made increasingly daring – and frequently undetected – incursions into Soviet undersea 'bastions' to further rattle the Soviet defenders.

It worked. What worried the Russians at the time – and has only become apparent since the end of the Cold War – was the increasing number of NATO, and especially US, incursions into what the Soviets felt were their defended air and sea space without being spotted. To an increasingly concerned Kremlin, the USSR's defences appeared to be full of holes. What this meant for deterrence and the nuclear balance was anyone's guess, but of one thing we can be sure: from the point of view of the Soviet leadership, the Cold War was suddenly beginning to look very unbalanced. As the Kremlin became aware of its vulnerability, the situation became more dangerous.

The year 1982 dawned with the Cold War raging anew and a positive blizzard of recriminations between East and West. The

CIA embarked on a vigorous funding campaign to destabilise the satellites of Eastern Europe. The KGB encouraged their own equally energetic campaign in the Western media, aimed at the peace movement and doing all that the Kremlin could to prevent the deployment of the Intermediate-range Nuclear Forces (INF), Pershing and Tomahawk cruise missiles, to their bases in Europe. Somehow the Soviet's own SS-20 theatre nuclear missiles were overlooked in the many vociferous denunciations of Reagan's 'crude warmongering' in Western newspapers.

Women marched on Greenham Common air base, the first cruise missile site in UK and set up camp, virtually imprisoning the facilities. The European press polarised sharply between left and right as denunciations flew between warmongers and peaceniks. A new Cold War arms race was accelerating and the propaganda campaign grew increasingly shrill on both sides. As early as autumn 1981, Andropov warned the Politburo that the United States was preparing a surprise nuclear attack on the USSR. The KGB and the GRU, he declared, must mount a single major joint intelligence collection effort, codenamed RYAN.

America's President Reagan and his advisors had been watching events carefully throughout the early years of his presidency. From the very beginning, Reagan had been prepared to negotiate with Brezhnev. This time, however, the American president intended the talking to be done from a position of clear US strength, not his predecessor Jimmy Carter's vacillating inconsistencies. That way, Washington knew it would get the Kremlin's full attention. After the old man's death in 1982, Reagan tried again with Brezhnev's equally geriatric successors, Andropov and Konstantin Chernenko. In 1983, Reagan wrote personally to the former that the best way forwards was for 'private and candid' personal communication between them both. Unfortunately, events intervened to stop all that.

One tragic incident of 1983 stopped any attempt at dialogue

and raised international tension to new levels. The shooting down of the defenceless Korean Air Lines Flight 007 over a Soviet closed military zone in the Far East by a Russian air defence interceptor was never going to endear the old men in the Kremlin to the free world's press, whatever the circumstances.

The episode shocked the world. *Time* magazine's cover for 12 September 1983, for instance, showed an artist's representation of a Soviet fighter flying away from the stricken airliner, with the headline 'Shooting to Kill'. The Soviet Union predictably denounced the flight as an American spying attempt, while the rest of the world saw the death of 269 innocent passengers as a ghastly blunder, caused by incompetent Soviet air defence controllers and trigger-happy Soviet pilots.

Denunciations flew in every direction. An indignant Washington played tapes of the conversations between the Soviet ground controller and the Sukhoi 15 interceptor to the press. The pilot's cold-blooded, matter-of-fact report that, 'The target is destroyed' sent a chill down even the most experienced Cold War spines. Either the Kremlin was deliberately sabre-rattling or the USSR's military was out of control.

However, we now know that all was not quite as it seemed. KAL Flight 007 had refuelled with an abnormally large fuel load for its scheduled flight plan. It was several hundred kilometres off course, but reporting back to air traffic control as if on its normal flight path. The pilot of the intercepting fighter jet had completed all agreed international procedures before finally opening fire. He had called the airliner on the international hailing frequency ('stud 16') before 'buzzing' it with his landing lights fully on, and had even fired a burst of highly coloured tracers across the Korean plane's nose as a warning. Despite all these warnings, the jumbo jet ploughed on. Captain Chun and his crew and 269 passengers paid the ultimate price as the aircraft plunged into the sea in Soviet restricted airspace. Many unanswered questions remain about KAL Flight 007.

That said, its effect on international relations in 1983 was disastrous, stoking up the Cold War tensions yet again, but now to a new peak. The USA banned Aeroflot airliners from landing in the US and the third – and final – major crisis of the long Cold War came slowly to the boil, with growing mutual suspicion and animosity on both sides.

Moscow was as alarmed as the White House at this unwelcome new direction in events. Foreign Minister Andrei Gromyko warned that 'The world situation is now slipping towards a very dangerous . . . precipice' and the Kremlin discussed Civil Defence and made discreet preparations for war. The Soviets appear, by 1983, to have been badly rattled by the new, tough line from Reagan and the accelerating pace of events. From MI6-SIS files, we now know that the Russian leadership genuinely believed that the USA was actively preparing for war, even a nuclear first strike. A wave of panic seems to have rippled through the Politburo.

The head of the East German Secret Intelligence Service, Marcus Wolf, was tasked by the KGB as a matter of urgency to recruit a high-level spy inside NATO headquarters in Brussels. Rainer Rupp, a German intelligence official posted to NATO HQ, and known as agent 'Topaz', was already spying for East Germany, also referred to at the time as the DDR (for Deutsche Demokratische Republik). He had delivered highly sensitive information from NATO to the Main Directorate for Reconnaissance (HVA) of the German Democratic Republic. He was ideally placed to know NATO's deepest secrets as his post at the time was head of the Current Intelligence Group, in the nerve centre of NATO.

He was specifically tasked with finding a copy of the Cosmic Top Secret NATO Plan to attack the USSR. The bewildered Rupp's report – that no such plan existed, and certainly not in the NATO duty officer's Cosmic Top Secret war safe (containing information that would be gravely damaging to

NATO if revealed) – was met with disbelief by the KGB, the GRU and the Kremlin. Such a plan must exist, insisted the panicking Russians: look again! Rupp was even given a special communications device to turn on the moment NATO took the decision to strike using its aggressive master plan to destroy the USSR. His assertion that really there was no such plan in NATO threw the Politburo and their KGB advisors into incomprehension and confusion. Of course there was a plan. There had to be. After all, the Soviets had one.

Once again the Kremlin leadership had made the fatal error of judging others by their own perverse and paranoiac standards. A bewildered Politburo was becoming increasingly jumpy and uncertain. A later, now declassified top-secret US Special National Intelligence Estimate (SNIE) from 1984 assessed that the US actions of the early 1980s had been deemed provocative by the USSR and were, in fact, contributing to the increased levels of Cold War tension. Marcus Wolf, the DDR spymaster, agreed, confirming this after the end of the Cold War and revealing that the KGB was obsessed by the dangers of a nuclear missile attack. He reveals that the Kremlin was 'in a panic' by 1982–83 and was insisting that all his East German agents should be now redirected to looking out for any warning of a nuclear strike and war.

Anti-Soviet sentiment in the West following the KAL Flight 007 incident was matched by a genuinely held view in Russia that America was re-arming and looking for some kind of provocation for war. The combination of major NATO exercises, the deployment of cruise and Pershing missiles and a worldwide US military alert following the killing of 241 US Marines in Beirut in 1983 convinced the Kremlin that America was preparing for a war that was now only a matter of time. Having walked out of the Arms Limitation Talks in Geneva, the Soviets were further hampered. By leaving the talks they had cut off their face-to-face dialogue with the West. In late 1983,

Soviet fears of an attack grew in direct relation to the Kremlin's growing fear of a US first nuclear strike.

Rainer Rupp confirmed this following his release from prison after his 1994 conviction for treason and spying:

> The Soviets were completely convinced that 'Able Archer' was the cover for a real nuclear strike. They believed that starting from this manoeuvre a strike aimed at decapitating the command, control and communi-cation centres of the Soviet army, the state apparatus and the party apparatus would be carried out with the help of the new ultra-modern and precise tactical nuclear missiles, Pershing II and cruise missiles for which you had a warning time of only five to eight minutes. [The Kremlin thought] . . . with these rockets, the criminal gang in the Pentagon hoped to decapitate the Soviet army, so that they – a quote that I myself have heard – 'would run around the farmhouse like a chicken with its head cut off'.

This was the climate in which the routine NATO exercise 'Able Archer' was launched in the autumn of 1983. For that year, the NATO exercise-planning staff had made some critical changes, including 170 radio-silent flights to airlift US troops to Europe, the use of updated nuclear weapon release procedures using new nuclear Command, Control, and Communications (C3) networks, the moving of NATO forces in Europe through each of the alert phases from DEFCON 5 to DEFCON 1, and the participation of political leaders including Margaret Thatcher, Helmut Kohl and Ronald Reagan. According to the KGB's Colonel Oleg Gordievsky, on seeing all this unfold, the Kremlin seems to have panicked.

A flash signal from Moscow Centre alerted all KGB residencies worldwide to go to 'intelligence of war alert' status.

The increased American activity, the loss of KAL 007 with its heavy loss of life, Andropov's apocalyptic warnings to the senior members of the Communist party over the previous two years and the Kremlin's own perception of vulnerability as the Americans rattled their cage with campaign of successful psychological operations suddenly all combined to create a crisis out of thin air. Even Ambassador Dobrynin, that unfailingly acute observer of the true state of the Cold War in Washington, admitted that 'both sides went a little crazy'.

According to Gordievsky, on the night of 8–9 November 1983 the Kremlin issued a 'Flash Top Secret' warning of war signal, claiming that US forces in Europe had gone on alert and that some units were being mobilised. This was just plain wrong, but in human affairs erroneous beliefs, however sincerely held, often encourage greater error. The Warsaw Pact was immediately ordered to war-readiness. They expected to be attacked; therefore they *would* be attacked. The Soviet High Command ordered a military alert in Eastern Europe. Ammunition was assembled at airfields and barracks, tanks brought to battle-readiness and leave cancelled. Soviet forces in Germany prepared to move to their war-time dispersal stations. Pilots were sent their target lists. The Kremlin talked about civil defence, a sure sign in the days of the Cold War that things were getting really serious.

Allied intelligence swiftly picked up this change in posture and alarmed NATO intelligence officers reported that 'the Soviets are up to something . . . ' Both sides now feared a preemptive attack, the Russians from newly deployed Pershing and cruise missiles, the NATO allies from a massive conventional invasion of Germany by 20,000 Russian tanks.

British agents deep inside the Kremlin and the Soviet High Command warned that the Kremlin really did believe that a nuclear war could be imminent. The American National Security Agency and British GCHQ signals intelligence intercepts confirmed KGB Colonel Gordievsky's explosive revelations. The Russians were very jumpy indeed and believed that NATO really was about to attack them.

The US and NATO leadership was astounded. The idea that the NATO Alliance was contemplating a nuclear first strike against Moscow would have been laughable, were it not so deadly serious. Most alarming of all, the fact that the Kremlin could actually believe such a thing to be even possible was a sinister insight into just how dangerous the Cold War had become, and how ignorant the Soviet leaders were. The experienced diplomat Anatoly Dobrynin was recruited one more time to help lower the temperature and convey messages.

The reassuring missives were promptly sent and 'Exercise Able Archer' swiftly wound down, much to the relief of the West Germans, who genuinely believed they were about to be invaded. It had been a frightening demonstration of just how quickly the nuclear pile at the core of the Cold War could go critical. Very few people in the West are aware of just how close the world came to war in that autumn of 1983. Seasoned Cold War observers believed that the crisis was every bit as dangerous and explosive as the Cuban Missile Crisis of two decades earlier.

Afterwards, a belligerent Andropov raised the stakes by refusing to talk to the US President, informing Reagan that the US deployment of cruise missiles had 'destroyed the very basis on which it was possible to seek an agreement' before retreating back into the comforting fastness of the Kremlin and the Politburo. Cruise missiles and Korean airliners meant that 1983 ended on a very sour note indeed.

President Reagan and Britain's Prime Minister Thatcher now realised that the time had come to make concerted efforts to connect with the collective leadership in Moscow, if only to reassure them. Reagan made conciliatory speeches. Thatcher asked for a meeting with the Kremlin's emerging new junior leadership. It was obvious to all that Andropov's successor, the ailing Chernenko – who assumed the reins of power in February 1984 – was not long for this world, let alone the premiership of the Soviet Union. Satirical television programmes in the West openly mocked Moscow's geriatric leaders, showing the Soviets pulling a succession of old men in wheelchairs out of a Kremlin deep freezer to answer Reagan's letters.

The result of the 'Able Archer' crisis was that by early 1984 both sides acknowledged that there was a genuine risk of things getting out of hand. A Cold War crisis always risked becoming a serious nuclear crisis. That was unacceptable. Each side now looked for new channels to keep a dialogue going and sought ways of keeping the lid on the kettle of what was looking like an increasingly dangerous Cold War confrontation. Behind the scenes, Foreign Minister Gromyko and US Secretary of State Schulz met in Stockholm for a long and serious private chat. Tensions eased and both sides relaxed.

The year 1984 ended with one of the more significant meetings of the Cold War. To the 'Kremlinologists' of Whitehall's Foreign and Commonwealth Office and the analysts of the British Defence Intelligence Staff, the identity of the most likely successor to Chernenko seemed fairly predictable. The Soviet leader was now so old and frail that he routinely handed over chairmanship of the Politburo's meetings to his head of the Secretariat, Mikhail Gorbachev. Despite the usual doubters and dissenting voices, Washington agreed with Whitehall's assessment of the USSR's likely choice, and in December 1984 the fifty-four-year-old Gorbachev and his glamorous wife Raisa came to London on an official visit, representing the Kremlin leadership. The visit was a great success, on all sides. The media liked the photogenic and cheerful Gorbachevs, so different from the usual Soviet visitors – miserable old men in ill-fitting suits. 'I like Mr Gorbachev...,' opined a smiling Mrs Thatcher, adding in a hugely significant judgement, 'we can do business together.'

Her snap opinion to a press query on the steps of Number 10 Downing Street was more important than she realised, for within three months Chernenko was dead, the third Kremlin leader to die in three years, and the fresh and relatively young Gorbachev had been unanimously elected chairman of the Politburo. The Kremlin's new leader knew better than anyone that he had to find a solution to the decades-old stand-off between East and West that was bankrupting the USSR. Reykjavik beckoned.

So, despite the nuclear nail-biting crisis of 1983's 'Exercise Able Archer', in the end it turned out to be a major catalyst in helping to end what we know as the Cold War. It was frightening in itself, but brought a lasting legacy of peace. It seems that military misunderstandings, especially the frightening nuclear ones, can sometimes lead to a solution.

In a curious way, and at the expense of badly frayed nerves – on both sides – it could be argued that nuclear weapons had served their purpose as a deterrent.

CHAPTER 19

SOME CIVILIAN NUCLEAR DISASTERS

'The International Nuclear and Radiological Event Scale is used for promptly and consistently communicating to the public the safety significance of events associated with sources of radiation.'
– THE INTERNATIONAL NUCLEAR AND RADIOLOGICAL EVENT SCALE USER'S MANUAL

While the military can often cover up the details about a nuclear accident under the cloak of 'national security', in the civilian world a nuclear mishap becomes news and public property very quickly. Everyone knows about the earthquake and tsunami that struck Japan on 11 March 2011, causing a calamitous accident at the Fukushima Daiichi nuclear power plant on the northeastern coast of Japan. You could hardly keep an event like that out of the headlines.

This disaster destroyed most of the Fukushima safety and power systems. The radiation released forced the evacuation of nearly half a million residents. This was one of the world's worst nuclear disasters and only the second catastrophe to measure Level 7 on the INES. It merits a chapter and a detailed exploration of what went wrong and why, all to itself.

Likewise with what is considered to have been the world's worst-ever nuclear disaster to date: Chernobyl. On 26 April 1986, an explosion and fire at this nuclear power plant near Pripyat – in what was then the Ukrainian Soviet Socialist Republic – destroyed a reactor and spewed massive amounts of radiation across the western Soviet Union and Europe. A quarter of a million people had to be evacuated and a large area remains dangerous and cordoned off to this day. Chernobyl is rated as the only other Level-7 INES.

The International Nuclear and Radiological Event Scale INES was established by the International Atomic Energy Agency (IAEA) and the Organization for Economic Co-operation and Development/Nuclear Energy Agency (OECD/NEA) and in 1992 all countries were recommended to formally adopt it. It categorises impact and severity. The IAEA defines such incidents as 'an event that has led to significant consequences to people, the environment or the facility'. The prime example of a 'major nuclear accident' is one in which a reactor core is damaged and significant amounts of radioactive isotopes are released at the risk of life and limb. Chernobyl is the worst example – so far.

Simplified, the seven levels of nuclear accident are shown here in ascending order of seriousness:

- When there are minor issues with a nuclear facility's safety components, but for the most part a considerable safety margin remains.
- When the radiation levels in a nuclear facility's operating area exceed 50 millisieverts (mSv) per hour, when a member of the public encounters radiation levels above 10 mSv, and when a worker is exposed to radiation levels over statutory annual limits.
- When a facility area is severely contaminated, causing non-

lethal injuries like radiation burns, but unlikely to result in high public exposure.
- When partial meltdown or fuel damage occurs, or when significant levels of radioactive material in an installation are released. With the exception of local food controls, no countermeasures are likely to be needed.
- When there is severe damage to the reactor core, and when large quantities of radioactive material are released within a site. In addition, when a limited release of material reaches the wider environment, and some planned countermeasures must be implemented.
- When a significant release of radioactive material is likely to need implementation of planned countermeasures.
- When there is a major release of radioactive material, which causes widespread and environmental harm, and which necessitates a considerable implementation of planned and extended countermeasures.

From the growing list of known nuclear incidents, we can assume that the question is not *if* a mishap will occur in future, but *when* and with what frequency? In *Normal Accidents: Living with High-Risk Technologies* (updated edition 2011), Charles Perrow explores how such mistakes happen. Perrow argues that the conventional engineering approach to ensuring safety-building, by introducing more warnings and safeguards, often fails because progressively more complex systems actually increase the possibility of a failure or accident. *Normal Accidents* introduced the idea of a single initial failure or mistake that starts a powder trail of disastrous, interconnected events, and observed that if just one element in the chain goes wrong, it creates the risk of a wider breakdown. Perrow's work provides a powerful framework for analysing risks and taking a hard look at the organisations that manage those risks.

His book was preceded by an equally important analysis

by British sociologist Barry Turner in his lesser-known, but similarly influential, book *Man-made Disasters* (1978). Turner's review of eighty major UK system failures reinforces Perrow's point and looks hard at the human factor in accidents. However, unlike Perrow, who concentrates more on the technological linkages and risks, Turner focuses in greater detail on the human factor in such events. His theory differs slightly from Perrow's – it gives a closer description of the organisational, management and communication failings that occur before an accident, the people involved and what they did.

Major accidents usually have a social and cultural backstory and a history of human interaction as well. For example, when a British European Airways (BEA) Trident jet crashed with the loss of 118 lives near London Heathrow Airport in 1972, it transpired that the pilots had failed to diagnose the fault amid at least nine other cockpit warnings and alarms that went off as a result. Yet the investigation also revealed that the personal relationships on the flight deck were poisonous. The younger crews of BEA wanted to go on strike for better conditions and more money. Stanley Key, a former RAF pilot who had served during World War II, considered this unprofessional for disciplined elite pilots. Captain Key was involved in a vicious personal quarrel in the crew room with his first officer just before take-off. Shortly afterwards, it stalled and crashed. To what extent was the accident caused by an uncorrected high tail stall and flat spin, and to what extent by poisonous personal relationships between the crew?

An interdisciplinary team from the Massachusetts Institute of Technology (MIT) studied Perrow and Turner's conclusions to find an estimate of how many nuclear accidents are statistically likely. They estimated that given the expected growth of nuclear power, at least four serious nuclear accidents would be expected in that period up to 2055. Since 1970, there have been five such incidents involving nuclear reactors going wrong and suffering core damage – Three Mile Island in 1979; Chernobyl in 1986; and three at Fukushima-Daiichi in 2011. That is an average of one serious nuclear accident every eight years worldwide.

It's important to remember that when nuclear power stations were constructed, it was on the understanding that they had a limited life expectancy. As Rob Smallwood, a nuclear engineer who worked on nuclear power stations for twenty years, points out:

> The accumulating factor on the power stations was always the build-up of radioactive nuclear waste. On a nuclear power station there is built a swimming pool twenty metres deep and filled with chemically pure water with a controlled specific gravity. This gives at least one metre of water to safeguard against the lethal gamma (γ) radiation. Sunk on the bottom of this pool is a robotic lathe. Periodically the engineers remove an isotope from the reactor and – all under water – skim a small amount off the outer shell. The isotope is then returned to the reactor. The residue, once tested for its radioactivity, is put into a lead-lined coffin and buried deep in the base of the power station. When this burial space is full, the station was meant to be closed down and the reactor decommissioned... The point is, they all had a limited shelf life...

This semi-experimental, short-term approach to civilian nuclear power plants certainly coloured the early days of atomic power. On 21 December 1951, the Idaho experimental plant – incorporating America's first nuclear power reactor, the EBR-I – began the first peaceful use of atomic energy, producing enough electricity to light a city.

In November 1955, scientists were investigating the reactor's responses by briefly cutting off the coolant flow and slowly turning up the power. The resultant rise in power was immediate and impressive. Power was switched off at once but by then the core had already started melting. As nuclear meltdowns go, the EBR-I incident was not that dramatic. Officially, it's not even classified as a serious nuclear accident. The scientists and engineers were experimenting. They just wanted to explore to the edge of knowledge, as curious scientists often do. But once the press caught wind of the core's unplanned meltdown – one of the first accidents associated with nuclear power – the accident was billed as an 'out of control' experiment by 'crazed scientists'.

The legacy of civilian nuclear accidents started on 12 December 1952, with a partial meltdown of the Canadian NRX reactor core at Chalk River in Ontario, resulting in the earliest major recorded civilian nuclear accident. The National Reactor Experiment at the Chalk River Laboratories was set up in 1944. It was the first nuclear reactor outside the United States to become operational, producing about one-third of the world's medical isotopes, as well as supplying secret shipments of plutonium for the US atomic bomb programme.

On 12 December, the reactor was set at low power with several of the cooling tubes routinely disconnected from the water-coolant system. The supervisor was in the basement and called up to the control room to tell his assistant to press two buttons to move the control rods to 'safe'. However, he gave the wrong numbers, providing instead the number of the button that withdrew the coolant safeguard bank. He immediately realised his mistake and shouted into the telephone to correct himself, but the assistant did not hear. He had put the telephone down to find and press the buttons.

After only ten seconds, the nuclear power output surged and the ordinary water in the temporary cooling system began to boil, as the power output quickly began to rise. The control room quickly spotted this and, puzzled and alarmed, began dumping the heavy water moderator. Inside the reactor, the sheathing around some fuel rods had burst. Air rushed in, causing a hydrogen-oxygen explosion. Fortunately, the radioactivity stayed inside the reactor vessel.

However, the water-cooling system was not shut down. This meant that the radioactive coolant water began leaking out into the reactor building. By 6:00 pm, the floor was awash; it was flooded to a depth of one metre within a few days, and had spread to the gas-holder's room and two rooms containing the heavy-water storage tanks. The first 650 litres of contaminated water was eventually pumped into a sterile tank and removed. To keep the remaining radioactive water running into the Ottawa River, a pipeline was constructed to redirect the contaminated water (which contained about 10,000 curies of long-lived fission products), to a sandy area about two kilometres away. There it was absorbed into the earth. Checks downstream found no evidence of radioactive contamination.

Decontamination and clean-up took many months. The USA helped; among the volunteer groups was a team of twenty-six volunteers from the US Navy under the command of future US President Jimmy Carter. The NRX reactor was out of service for two years. Fortunately, there were no fatalities or serious injuries, although some personnel had been exposed to temporarily high radiation levels during the incident.

This meltdown was followed by a second accident at Chalk River, in 1958, involving a fire in the nearby National Research Universal (NRU) reactor building. A robotic crane accidentally pulled a hot uranium rod out of the reactor vessel. The uranium caught fire and the rod snapped, its largest segment falling into the containment vessel. The valves of the ventilation system were opened, and a large area outside the building was contaminated. Efforts to extinguish the blaze saw scientists and maintenance men sprinting along the holed vessel like Keystone Cops and hurling in bucketfuls of wet sand.

On 5 December 2008, there was yet another serious radioactive leakage at Chalk River, an escape of heavy water that contained tritium, although it was successfully controlled. However, in mid-May 2009, the leak returned at a greater rate, after which the plant was closed for more than a year, contributing to a global shortfall in medical isotopes.

The catalogue of nuclear accidents goes on and is often shrouded in scientific amnesia. The breakdown of the Enrico Fermi Unit at Frenchtown Charter Township, Michigan, on 5 October 1966, is little known. It was the USA's first and only commercially operating liquid metal fast breeder reactor. A coolant flow blockage to the nuclear fuel led to the partial meltdown of two fuel assemblies. That caused a component within the reactor vessel to loosen, which forced the radioactive liquid to find an alternative path. The danger only became apparent when temperature alarms went off. The core began to heat up and fuel rods melted on reaching temperatures approaching 400°C. The Fermi reactor was shut down for repairs, but was never fully operational again and was decommissioned in 1975. But the lessons learned from this failure in the nuclear cooling system were not passed on to other users – with catastrophic consequences.

One area where nuclear secrets were never shared was behind what was known, during the Cold War era, as the 'Iron Curtain'. We have already discussed some of the Soviets' blunders and accidents, but in 1977 a near nuclear disaster occurred in Moscow's satellite state, Czechoslovakia.

The decision to build a nuclear power plant at Jaslovské Bohunice in Czechoslovakia was made in 1956. It took another sixteen years before the power station was fired up on 24 October 1972. The facility was intended to use natural uranium mined in Czechoslovakia as the basic fuel. Because of its experimental design, however, it suffered more than thirty unplanned shutdowns caused by series of accidents and malfunctions. On 5

January 1976, there was a fatal accident. Two workers were killed owing to a leak of carbon dioxide coolant gas. In true Soviet style, these accidents were concealed, although wild stories circulated among the public.

On 22 February 1977, the country's most severe nuclear incident occurred, as a result of design faults and human error, when fuel rods were replaced while the reactor was kept running. The operators mistakenly failed to remove some silica gel pellets that had fallen into a new fuel element from a damaged pack. This caused overheating of the fuel and heavy water came into direct contact with the coolant. The dropped pellets blocked up the coolant flow, which caused the fuel to begin overheating; this, in turn, forced the contaminated heavy water into the reactor's gas circuit. Parts of the secondary circuit then began to catch fire and had to be 'scrammed' to prevent a disaster.

There was a serious leakage of radioactivity – naturally, hushed up by the Communist authorities – and contamination of the area was widespread, with radioactive caesium and other isotopes detected for sixteen kilometres downstream of the reactor and the associated riverbanks and sediments of Manivier Canal and Dudvah River. The Czech investigation concluded that:

> Contamination of the environment by radionuclides from accidental A-1 NPP needs permanent monitoring of underground water, sediments, soils and aerosols and consistent realisation of measures for remediation and isolation of mostly contaminated soils as well as permanent outflow of underground water in neighbourhood of area of the A-1 NPP.

Despite the accident being only rated Level 4 on the INES, in May 1979 the Czech government decided to decommission and decontaminate the entire plant.

A far more notorious accident involving coolant took place at the Three Mile Island Nuclear Power Station in the USA. This civilian power station was located on an island in the Susquehanna River near Harrisburg, Pennsylvania. The plant was built in 1974 on a sandbar sixteen kilometres downstream from the state capital in Harrisburg and run by electricity company Metropolitan Edison. In 1978, a second reactor began operating on Three Mile Island to supply power to large parts of Pennsylvania. The two separate units, TMI-1 and TMI-2, used water from the river to cool the intense heat produced by 150 tons of uranium.

At 4:00 am on 28 March 1979, the worst accident in the history of the US nuclear power industry began when a pressure valve in the Unit-2 reactor at Three Mile Island failed to close. Cooling water, contaminated with radiation, drained from the open valve into adjoining buildings, and the core began to overheat. For five days, authorities struggled desperately to prevent the crippled plant from spewing dangerous quantities of radiation into the environment.

After the cooling water began to drain out of the broken pressure valve, emergency cooling pumps automatically went into operation. But the control room operators misunderstood their instrument readings and made the situation worse by shutting off the emergency water system. Without cooling water, the operators faced a runaway core chain reaction. By early morning, the core had heated to more than 4,000°C, just 1,000 degrees short of meltdown.

News of the accident leaked to the outside world. Metropolitan Edison's PR team and lawyers tried to talk their way out of the growing crisis by claiming that 'no radiation had been detected outside the plant'. Unfortunately for the corporate suits, alarmed environmental inspectors were already warning of increased levels of radiation nearby as a result of the contaminated water leak, and Pennsylvania's state governor was considering calling an evacuation.

After the radiation leak was discovered on 30 March, residents were advised to stay indoors. Experts were uncertain whether the build-up of hydrogen gas would fuel a further meltdown or possibly a giant explosion. As a precaution, state governor Dick Thornburgh advised 'pregnant women and pre-school-age children to leave the area within a eight-kilometre radius of the Three Mile Island facility until further notice'. Despite Metropolitan Edison's bland assurances, within days more than one hundred thousand people had stampeded to flee the surrounding towns.

By early evening, the harassed and panicky plant operators had worked out for themselves that they needed to get water moving through the core again and restarted the cooling pumps. The temperature began to drop, and pressure in the reactor was reduced. Later, it was estimated that the nuclear reactor had come within less than an hour of a complete meltdown and almost certainly an explosion that would have released tonnes of radioactive gas into the atmosphere. On the day, although more than half the core had melted, it had not broken the containment vessel, and no radiation was escaping. It had been a lucky escape.

Ultimately, the Three Mile Island accident was a matter of great concern, but it did not lead to a public health disaster, as the pressure vessel and power plant's structure stayed strong, with only a relatively small volume of lethal radiation ending up in the atmosphere.

On 1 April, President Jimmy Carter came to Three Mile Island to be seen inspecting the plant. Carter was a trained nuclear engineer, and – as we've already seen – had helped dismantle a damaged Canadian nuclear reactor while serving in the US Navy. His visit calmed and reassured local residents and the nation. And slowly, the reactor cooled.

The subsequent enquiries discovered some unpleasant truths. The reactor operators had not been trained in accidents.

No real emergency operating procedures were found and the Nuclear Regulatory Commission (NRC) had neglected to set up formal lines of communication with the utility companies they were supposed to be regulating. Human error had made a bad technical situation much worse. Workers in the control room faced a situation for which they had not been prepared, and were forced to make their own decisions as events unfolded. They were not helped by faulty instrumentation in their control room.

At the height of the crisis, plant workers were exposed to unhealthy levels of radiation, although no one outside Three Mile Island had their health adversely affected by the accident. The undamaged reactor at Three Mile Island was shut down during the crisis and did not resume operation until 1985. Clean-up continued on Unit-2 until 1990, but it proved too damaged to be made usable again.

It took 11 years and about $1 billion to clean up TMI-2. During those operations it was discovered in 1985 that large portions of the reactor core had melted and fused. Three Mile Island had an impact out of all proportion to its real damage and casualties. The incident greatly eroded the public's faith in nuclear power. For the next three decades, not a single new nuclear power plant was ordered in the United States. In 2010, President Obama finally suggested Congress reconsider nuclear power and proposed $8 billion in loans to begin building new plants in Georgia.

It would be only too easy to go through a lengthy litany of other nuclear accidents to make the point that nuclear power is potentially lethal once it escapes its bonds. A few other selected examples will suffice to draw attention to the continuing widespread danger of even peaceful civilian nuclear power. On 17 October 1969, something went badly wrong at the Lucens nuclear power reactor in Switzerland. It was an experimental 6-megawatt, Swiss-designed pilot nuclear plant; for safety reasons, the reactor – moderated by heavy water and cooled by carbon dioxide – was built in a cave at Lucens in Vaud. Construction began in 1962 and the power station went on line in December 1966. It was fuelled by enriched uranium, alloyed with chromium cased in magnesium alloy, and inserted into a graphite matrix.

The initial plan was to run the plant until the end of 1969, but during start-up on 21 January 1969, a loss of coolant resulted in a runaway reactor. Corrosion of some of the magnesium alloy fuel element components blocked one of the fuel channels and the lack of coolant caused cladding to melt and further block the channel. The increase in temperature and exposure of the uranium metal fuel to the coolant eventually caused the overheating fuel to catch fire and burst a tube, which allowed radioactive cooling gas to leak out of the reactor. Alarmed workers shut down the pile and fled.

Although the cavern that housed the reactor had to be sealed off due to serious contamination, workers were not physically affected by the accident. While the cavern was decontaminated and the reactor taken apart in the years that followed, it took until 2003 for all radioactive waste to be removed.

Switzerland's only nuclear accident was assessed as a Level 4 on the INES, categorised as an 'accident with local consequences'. But, yet again, it was a clear warning that when things go wrong with nuclear devices, they often go seriously wrong.

Later events in France and Scotland confirmed the dangers. On 17 October 1969 at the Saint-Laurent plant on the River Loire, uranium in the reactor core melted, bringing about a partial meltdown and generating high levels of radioactivity. The clean-up and repairs took a full year to complete. On 13 March 1980, things went wrong again here, when the graphite in one of the reactors fused with the core, causing overheating. The reactor began to run away and had to be shut down in an emergency scram. This was also classified as a Level-4 incident on the

INES, and has been described as the worst nuclear accident in France. However, Andrá Gauvenet, the Inspector General for Nuclear Safety and Protection, soothed French fears by stating that there had been no casualties and no release of radiation.

Nonetheless, over the years the Saint-Laurent plant ran up a history of near accidents. On the morning of 12 January 1987, ice clogged the water going into the central reactor due to a hard frost in the Loire, and resulted in the loss of normal cooling. The army was quickly called in to use explosives and blast away the ice blocking the coolant water intakes and the reactor was returned to normality.

The catalogue of French nuclear accidents goes on. In December 1999, a massive storm provoked the partial flooding of some reactors at Electricité de France's (EDF) Blayais plant in the southwest of the country. Critics of nuclear power argued that the flooding nearly caused a major catastrophe, because it briefly cut off power at the plant. On 12 May 2004, radioactive sodium was released into the atmosphere during a leak test of new steam generators at of one of the reactors at the B plant. The incident 'was of no consequence for the environment' according to EDF.

Areva's Socatri site in southeastern France houses four nuclear reactors. In July 2008, a total of 30 cubic metres of a liquid containing natural uranium was accidentally poured on to the ground and into a river while the tank was being cleaned. The pure uranium was much less dangerous than enriched uranium, but France's ASN nuclear watchdog rebuked Areva for mishandling the accident.

In November 2009, a radioactive fuel assembly rod got stuck in the pressure vessel at EDF's Tricastin plant in southeast France, raising the risk of an accident. A similar incident had taken place in September 2008 in the same reactor during refuelling operations. It took two months for engineers to stabilise the position of the rod and organise its unhooking and removal.

France's nuclear problems continued. In June 2011, a minor and fairly common incident that involved internal leakage at EDF's Paluel 3 nuclear reactor was reported by French investigative website Mediapart. This was followed by an incident at the Marcoule nuclear site on 12 September 2011. One person was killed and four were injured – one with serious burns – after an explosion in a furnace used to melt down nuclear waste and recycle it for energy. No radiation leaks or damage to the plant were reported.

By any standards, this list of French examples is a story of continuing nuclear problems, little known to the outside world, but they all spell danger. The French are now keenly aware of the dangers of nuclear energy. A 2019 study concluded that a major disaster damaging one of France's fifty-eight nuclear reactors and contaminating the environment with radioactive material would displace an estimated one hundred thousand people, destroy crops, wreck tourism for years and create massive power outages. 'A major accident would have terrible consequences,' admitted Jacques Repussard, the director general of France's public-funded Institut de Radioprotection et de Sûreté Nucléaire, in 2013, 'but we would have to deal with them because the country wouldn't be annihilated, so we have to talk about it, however difficult it is.'

Britons have no cause to feel smug about France's nuclear problems. Behind the duplicities and half-truths of Whitehall, a similar tale of hidden dangers emerges, perhaps best summed up by the little-known history of what has been called Britain's nuclear secret site – Chapelcross.

This former nuclear power station lies near Annan in Dumfries and Galloway on the site of a remote ex-RAF airfield. One of the oldest in Britain, it became operational in February 1959 and was the first such facility in Scotland. Its four 50-megawatt electrical Magnox reactors were designed to perform two functions – supply weapons-grade plutonium for the military and produce cheap electricity for the local civilian power grid.

Chapelcross's role in producing plutonium for thermonuclear bombs has not been made public, for obvious security reasons. The British nuclear weapons programme began during the 1940s, and production of British nuclear weapons started in 1950 using material produced at Windscale. However, after the Windscale fire in 1958, production shifted to Calder Hall and Chapelcross which, unknown to many Scots, served as one of the main sources of weapons-grade plutonium until as late as 1998.

The plant's nuclear reactors also produced tritium gas, an essential component to boost the yield of nuclear weapons. An estimated 5,500 tons of Magnox depleted uranium was also held in storage at Chapelcross, part of a military stockpile formerly intended for use in depleted uranium weapons.

Chapelcross was not trouble-free. In 1967 there was a partial meltdown in reactor No. 2, when a fuel rod broke. The reactor was closed for two years and restarted in 1969. The press was told that no radioactivity had been released. Later, the discovery of a forty-year-old hairline crack in one of the sixteen heat exchangers resulted in a six-month shutdown.

In 2001, Chapelcross hit the news again. A basket of radioactive spent fuel elements was dropped and actually remained undetected for several days. An in-house investigation found that, apart from the delay in detecting this occurrence, there was no hazard to either plant operators or the public. This bland explanation was not good enough for the local Member of Parliament, who tabled a question in the House of Commons demanding to know what was going on:

> Tabled 18 July 2001. That this House notes the widespread concern throughout Scotland at the accident at the Chapelcross Nuclear Power Station; is extremely concerned that twelve fuel rods apparently remain unaccounted for and that there appear to have been attempts to cover up

the incident; and calls upon the Government to carry out a full and independent inquiry into this potentially catastrophic accident.

While the civil service could easily bat away intrusions from nosy MPs, the questions in the House brought unwanted attention to a very sensitive site and a very delicate subject: British nuclear safety. Power production at the Chapelcross plant ended in 2004. In February 2013, the last of more than 38,000 fuel elements was removed from the decommissioned nuclear plant. In January 2017, plans were approved to increase the size of a radioactive waste storage facility, able to store intermediate level waste for up to 150 years, awaiting a final disposal location available off-site. The problem of nuclear waste increases in severity as the years march by.

Chapelcross is merely another in a variety of civilian nuclear accidents. There are simply far too many for us to examine them all in detail, as the list of all reported nuclear INES issued by the International Atomic Energy Agency (IAEA). They all serve to show just how vulnerable and dangerous even civilian nuclear plants can be. While not the most dangerous, civilian nuclear plants are statistically the most likely source of nuclear accidents. Chernobyl was the worst.

CHAPTER 20

CHERNOBYL - THE WORST SO FAR

'In major commercial reactor accidents, there always seems to be a single operator action that starts the downward spiral into irrecoverable disaster.'
– JAMES MAHAFFEY, *ATOMIC ACCIDENTS* (2014)

The enormous cost of the disaster at the Chernobyl Nuclear Power Plant on 26 April 1986 had widespread consequences. It ruined the reformist ambitions of President Gorbachev and forced him, and the Central Committee of the USSR, to confront the stark reality of the country's collapsing command economy. And it could have been avoided. In a conference later that year to discuss the lessons of the incident, Western scientists revealed that the nuclear explosion stemmed largely from reactor design defects that the British had warned the Soviets about in 1977. It seems that Chernobyl was an accident waiting to happen.

During the 1977 conference, Soviet scientists recognised that the catastrophe was partly caused by design faults in the power plant's RBMK reactor, rather than by those who operated it. Reactor designers present from the West agreed with this conclusion; to quote Dr Adolph Birkhofer, who headed the West German Reactor Safety Commision: 'We don't agree that this accident was caused more by human error than technical deficiencies. That reactor was too hard to operate.'

'Several facets of the Russian reactor were wholly unsatisfactory,' added the chairman of the UK's Central Electricity Generating Board, Lord Walter Marshall. 'The Soviet Union chose its design because they thought it would save money. This was a judgement made in advance. They were warned back in 1977.' Back then, continued Lord Marshall, the British had spotted these flaws when discussing the possibility of building their own similarly designed reactors. He listed seven specific problems with the Soviet reactors:

- A 'positive void coefficient', meaning that when the reactor loses cooling water, the nuclear reaction accelerates.
- Whereas Western models of reactors used only one type of control rod, the Soviets used four or five types, causing a wide variation of neutron levels, and making the reactor much harder to control.
- 'The Soviet Graphite Core is hot and a fire risk in an accident', so when Chernobyl's water tubes ruptured, the surrounding hot graphite is likely to have been a contributory factor in the steam explosion that ensued.
- Due to the Soviet reactor's weak structure, the smallest pipe rupture could jam the whole system, which was indeed what happened at Chernobyl in April 1986.
- Insufficient concrete and steel surrounding the reactor, making it vulnerable in the event of a serious accident.
- With no mechanism present for regular cooling of fuel in the RBMK, a continuous safety system was necessary to cool the fuel by water spray.
- Lastly, a complicated piping system – dubbed 'a

plumber's nightmare' by operators in the West – made it less likely to spot any weaknesses or cracks.

This flawed and potentially dangerous reactor was to be the cause of the nuclear disaster at Chernobyl, which was located in Ukraine, a republic of the Soviet Union at that time. It is now fairly clear what happened on that night in 1986. The Soviet-designed RBMK nuclear reactor dates back to the 1970s. There were almost twenty of these reactors completed. As the disaster showed, and the 1986 conference revealed, the RBMK's serious design flaws were dangerous. In particular, the location of the control rods, the containment structure and the reactor's 'void coefficient' proved to be quite unsafe. This last is important because the RBMK design has, in the jargon of nuclear engineers, a 'void coefficient', which means a distressing tendency to boil too many empty bubbles in the reactor coolant. That night the technicians got it wrong at Chernobyl, the reactor boiled over and the fuel caught fire, just like a gigantic, out-of-control kettle boiling dry.

In the early morning of 26 April 1986, reactor No. 4 of the Chernobyl nuclear station exploded, bringing about what the United Nations has called 'the greatest environmental catastrophe in the history of humanity'. It exploded because unqualified and untrained control-room technicians were trying to carry out (of all things) a safety test. Grigori Medvedev was scathing about the events in his 1991 book, *The Truth About Chernobyl*, a treasure trove of primary sources and first-hand evidence of what really happened on the day. Having served as deputy chief engineer at No. 1 reactor unit of the Chernobyl Nuclear Power Plant in the 1970s, Medvedev knew the facility well and was sent back as a special investigator immediately after the 1986 catastrophe. Secretive Communist nuclear bureaucrats must have winced as they read his excoriating revelations:

The mere fact that the operators were carrying out an experiment that had never been approved by higher officials indicates that something was wrong with the chain of command. The State Committee on Safety in the Atomic Power Industry is permanently represented at the Chernobyl station. Yet the engineers and experts in that office were not informed about the programme. In part, the tragedy was the product of administrative anarchy or the attempt to keep everything secret.

What happened at Chernobyl was a chain of human mistakes leading to a catastrophic accident. RBMK reactors such as those at Chernobyl rely on huge quantities of water as a coolant. Reactor No. 4 at Chernobyl included about 1,600 individual fuel channels, each of which required coolant flow of twenty-eight tonnes of water every hour. Chernobyl's reactors had three back-up diesel generators to ensure cooling even if the electricity to the main pumps failed. However, it took about a minute for them to spool up to provide emergency power. This delay was considered a potential safety risk and required checks.

There had been three previous tests, in 1982, 1984 and 1985, to make sure that the diesel generators could keep the reactor coolant flowing in a power cut; all had proved flawed. The 1986 test called for temporarily switching off the emergency core cooling system, and was given a special approval from the chief nuclear engineer. Regarded as a purely electrical test, not a nuclear check, it was to be conducted during the day shift of 25 April 1986 as part of a carefully planned reactor shut down.

While the day-shift team had been informed of the operating conditions for testing the reactor, once the correct conditions were reached, another team of electrical engineers was brought in to carry out the one-minute test of the new voltage regulating system.

After completing several other, unconnected maintenance tasks, the day shift team got ready to begin their experimental test at 2:15 in the afternoon. The necessary steps for the temporary switching off of the emergency core cooling system were activated. However, a problem elsewhere intervened. The power station at Kiev unexpectedly went offline and at 2:00 pm the Kiev electrical grid controller called for Chernobyl to keep supplying power as electricity was needed to satisfy demand when it peaked that evening. Chernobyl's obliging plant director complied.

At 4:00 pm, the Chernobyl day shift was replaced by the evening shift. At just after 11:00 pm, Kiev's grid controller allowed the reactor shutdown to resume. This would have grave consequences, as the incoming night shift team (previously only expected to maintain decaying heat cooling systems) did not have any prior knowledge of, or training in, carrying out the experimental test, a test they were now obliged to prepare for in a very short time. Supervising and directing the exercise would be Chernobyl's deputy chief engineer, Anatoly Dyatlov. The test – a standard check of the electric emergency power safety back-up – required reactor No. 4's nuclear power output to be decreased until it reached a level of between 700 and 1,000 megawatts. At just after midnight, an output of 720 megawatts was duly reached.

Soon the reactor power would decrease to around 500 megawatts, with the reactor control now in manual control, and it was here that the power accidentally entered a state of near-shutdown. Why and how this happened cannot be established for certain as both Akimov and his assistant nuclear engineer Toptunov died the following month, but both human error on the part of the latter and equipment failure have both been cited as reasons for the unexplained power drop.

We know that an output of 720 megawatts was reached at 12:05 am on 26 April. By now, the reactor was producing just

5 per cent of the initial power level required for the test. The operation of the RMBK reactor at the low power level was accompanied by unstable core temperatures and coolant flow, and by accelerating evidence of increased neutron movement among the growing bubbles, triggering several alarms. We know that the increasingly confused control-room operators were receiving repeated emergency signals about the low levels in the steam-water separator drums, and large variations in the flow rate of feed water, as well as from the neutron power controller. There were emergency alarm signals between 12:35 and 12:45 am, but these were not acted on, reportedly in order to save the low reactor power level for the test.

The test finally began at 4 seconds after 1:23 am with a shutting-off of the steam to the turbines, resulting in the turbine generator running down. Everyone waited for the diesel generators to kick in, but the slowing-down of the turbine generator led to a reduction in the power it was producing for the cooling pumps. The rate of water flowing to cool the reactor slowed, leading to more and more voids (bubbles/steam forming within the coolant), which intensified the chain reaction. This all happened in seconds, and now the core power of reactor No. 4 was in danger of accelerating uncontrollably.

The local automatic control system would have successfully compensated for much of the experiment by inserting control rods into the reactor core, which would serve to restrict the rise in power. But this system controlled only a dozen rods, as most of the others had been manually retracted earlier by the day-shift reactor operators as preparation for the test. The night-shift operators were perplexed and began asking each other why the reactor was not damping down automatically.

At 1:23 am and 40 seconds, as the experiment closed down, a pre-planned scram of the reactor kicked off. The reactor emergency protection system was engaged to re-insert all the graphite control rods.

Something went wrong. The emergency scram initially increased the reaction rate in the lower part of the core. A few seconds into the scram, a power spike occurred. The core overheated immediately. Nuclear reactions happen very quickly. Some of the fuel rods broke because of the heat and jammed the control rods. The reactor output shot up. The power spike increased the temperature even more, escalating the steam pressure. The safety cladding on the fuel broke away; hot fuel elements entered the coolant, bursting and blocking the operating channels. The reactor was out of control.

Fuelled by a raw nuclear reaction, the temperature shot up and there was a devastating explosion, which destroyed the reactor casing and the upper biological shield. The explosion that followed, the first to be heard by many, ruptured more fuel elements, and severed most of the coolant lines which were feeding the reactor chamber. The coolant became highly pressured steam, which blew out the reactor core About five seconds later, a second, more powerful explosion occurred, effectively cutting off a nuclear chain reaction by blowing everything apart.

Eyewitnesses outside Unit 4 saw burning objects and sparks flying into the air, which then started a fire on the roof of the machine hall. One-quarter of the hot graphite blocks and overheated material from the fuel channels was ejected, and this material was blown out of the reactor building. Outside air flowed into the core, feeding the fire and raising the high temperature inside. This air, coming into contact with the hot graphite, started a second graphite fire.

This explosion also compromised more of the reactor containment vessel and ejected hot lumps of the graphite moderator. The ejected burning graphite and the demolished channels caught fire, spreading radioactive fallout and the contamination of the whole region downwind.

A survivor, Alexander Yuvchenko, witnessed the building being destroyed by the second explosion, and would later describe the striking sight of a beautiful shaft of blue light 'flooding up into infinity'.

The second explosion, which produced the majority of the damage, has been estimated as equivalent to about ten tonnes of TNT. A full nuclear chain reaction was only prevented because the explosion scattered the reactor core. Many nuclear scientists believe that what happened at Chernobyl was nothing less than an atomic bomb 'fizzle' reaction (see page 5) and could have been much worse if the core had not been blown into pieces.

It all happened very quickly. One possible explanation came from Lars-Erik De Geer, an expert in nuclear forensics formerly of the Swedish Defence Research Agency, and his colleagues in 2017. De Geer believes it was the other way round, that it was a *nuclear* explosion that came first:

> We believe that thermal neutron nuclear explosions at the bottom of a number of fuel channels in the reactor caused a jet of debris to shoot upwards through the refuelling tubes. This jet then rammed the tubes' 350-kilogram plugs, continued through the roof ... The steam explosion which ruptured the reactor vessel occurred some 2.7 seconds later.

De Geer is convinced that the Chernobyl disaster was effectively the first part of a nuclear explosion. He points to the fact that the amount of radiation released is estimated as being at least one hundred times more powerful than the radiation released by the atom bombs dropped on Hiroshima and Nagasaki.

Whatever the cause, at Chernobyl on 26 April 1986, something akin to a small nuclear bomb exploded. A plume of hot radioactive gases was borne away on the wind. The fallout, extremely uneven because of the shifting wind patterns, extended 1,900–2,100 kilometres from the point of the accident. Roughly 5 per cent of the reactor fuel – or 7 tons of fuel containing 50–100 million curies – was released into the atmosphere.

First to spot the danger were the Swedes.

The alarm was sounded at Forsmark, Sweden's second largest nuclear power plant, early in the morning of 28 April 1986. One of the employees went through a radiation monitor and a warning sound went off. The source was his shoes, which showed high levels of radiation. The Swedes were mystified. It was not immediately clear where the radioactive material had been released, despite staff scanning all their radiation detection instruments. 'We found nothing,' said Claes-Göran Runermark, then the operation manager. 'We went over all the radiation detection systems over and over again, and there was nothing from Forsmark.'

Eventually, the Swedes identified the contamination on the worker's shoes as coming from the wet grass outside. On examination, they realised that the source of the radioactive particles had to be a Soviet nuclear power station. Also, during the weekend the wind had blown from the southeast and it had rained in the northeastern parts of Sweden, depositing radioactive fallout on the ground in that area. All the evidence was pointing towards contamination from the Soviet Union. Questioned late that evening, nearly three days after the disaster, the Russian authorities admitted that there had been an accident 1,100 kilometres away in the Ukrainian town of Chernobyl.

By 4 May 1986, radioactive fallout containing the deadly isotope caesium-137 and other nuclear contamination covered an enormous area, including Byelorussia (now Belarus), Latvia, Lithuania, the central portion of the Soviet Union, the Scandinavian countries, the Ukraine, Poland, Austria, Czechoslovakia, Germany, Switzerland, northern Italy, eastern France, Romania, Bulgaria, Greece, Yugoslavia, the Netherlands and the United Kingdom.

Estimates of the effects of this fallout range from 25,000 to 100,000 deaths from cancer and genetic defects within the following 50 years. In particular, livestock in high rainfall areas received unacceptable dosages of radiation.

The Soviet Air Force was ordered to seed rain clouds gathering over Byelorussia (Belarus) and black-coloured 'dirty' rain was reported falling at the city of Gomel. Byelorussia seems to have received about 60 per cent of the contamination that fell on the Soviet Union, but radioactive fallout also fell widely across northwestern Ukraine. It is now estimated that about one million people could have been affected by radiation in surrounding countries. Further afield, the Chernobyl accident spread radioactivity to the Alps, the Welsh mountains and the Scottish Highlands. The resulting contamination caused groundwater to carry residual radioactivity downstream, especially in Scandinavia. The downwind hazard was even reported in Ireland, 2,500 kilometres to the west. Radioactive fallout knows no borders.

Back in Ukraine, once the Soviets had realised that the cat was out of the bag over the disaster, something akin to panic seems to have set in at governmental level. According to journalist Kim Willsher:

> In their desperation to save face, the Soviets were willing to sacrifice any number of men, women and children. Even as radiation spewed out of the plant from the burning reactor core, local people told John and me [Willsher and her photographer, John Downing] how they had seen Communist apparatchiks in the area spirit their families to safety in Moscow while the residents were being urged to carry on as if nothing had happened. In Pripyat, the satellite city built for Chernobyl workers, windows were left open, children played outside, and gardeners dug their allotments.

Residents were told that the plume of deadly radioactive dust was, in Willsher's words, 'just a harmless steam discharge'. Eventually, the ruling Communist Party ordered a mass evacuation of the city of Pripyat two days after the explosion and 50,000 people fled, by which time some were already showing signs of radiation sickness. A thirty-kilometre exclusion zone was slapped round the town.

The Communist authorities continued to lie to their own people. On the Soviet TV news on 29 April, more than three days after the catastrophe, with the reactor fire still burning, a female presenter announced, as she came to item six on the programme, 'There has been an accident . . . Two people have died.' People, schoolchildren among them, all over the Soviet Union, including Byelorussia and Ukraine, were encouraged to take part in the customary parades and festivities of May Day, despite the radioactive particles in the air.

In the days and months after the world's worst nuclear accident, the Kremlin kept to its usual default position on Soviet disasters. It dissembled, denied, deceived and lied. Few were fooled. In the USA, the CIA and the National Reconnaissance Office published embarrassing satellite photographs showing the damage, and the media kept asking awkward questions. To this day, Moscow's campaign of denial remains unchanged. A Soviet embassy official stated in 1986, 'The problem is getting better. It is not out of hand. It is improving. But unfortunately, it is not over yet.' As the *Washington Post* wrote in 2016, 'Compounding the catastrophe was the lack of transparency from the Soviet Union. In the days that followed, Soviet officials refused to elaborate on the extent of the disaster.' But Chernobyl was simply too big to hide or cover up.

Cleaning the mess up was obviously the most urgent task. The plant explosion had not only released a lot of radiation, it had also covered a vast area, contaminating some 200,000 square kilometres of Ukraine. At least 20 per cent of Chernobyl's 190 tonnes of uranium had been sent into the atmosphere and 240,000 recovery workers were called in. Over the next two years, thousands more were recruited for the clean-up, according to the World Health Organisation. The Soviet authorities issued no less than 600,000 certificates to the 'liquidators', as these workers were termed.

First on the scene were the dazed and shocked survivors of the explosion at the plant. They were swiftly reinforced by local firefighters who rushed to the accident, followed by civil defence troops from the Soviet Armed Forces. Most of these first responders worked without specialist radiation safety clothing as they struggled to put the fires out and remove contaminated materials. Special Internal Troops and police sealed the perimeter of the plant, controlling security, access and population evacuation.

Soviet Air Force and civil aviation units were mobilised to 'bomb' the open wreckage of reactor No. 4 to put out and cover up the fires and radiation. Lethal levels of radiation around the reactor prevented ground liquidator teams from getting too close to the hot areas.

Helicopters could reach and leave the area above the building and were tasked with dropping sandbags and lead slabs to cover the damage, photograph the site and monitor radiation. One of the pilots, Mykola Melnyk, flew forty-six sorties over the radioactive reactor building, and was later honoured with a Hero of the Soviet Union award for hovering and placing a giant eighteen-metre radiation probe in the building at the third attempt.

Although Soviet officials initially put the number of fatalities at a ludicrous thirty-one, the United Nations estimates more than 3.5 million people overall were affected one way or another. While there is rough agreement that a total of either thirty-one or fifty-four people died from blast trauma or acute radiation syndrome as a direct result of the disaster, it is well-nigh impossible to track the total number of Chernobyl casualties. Average radiation dosages varied from 12 rads to 50 rads (more than 100 rads a day induces radiation sickness), although of the liquidators who died within a month, some had received a horrifying 6,000 rads.

The accurate long-term health effects and number of deaths due to the disaster vary widely, with estimates ranging from 4,000 to no less than 93,000. Deaths thirty years on are now impossible to identify in the wider mortality figures. The Chernobyl Union, the liquidators veterans group, claims that, '25,000 of the Russian liquidators are dead and 70,000 disabled, about the same in Ukraine, and 10,000 dead in Belarus and 25,000 disabled,' which makes a total of around 60,000 dead and 165,000 disabled. Again, we don't know exactly and these figures cannot be verified.

In the months after the explosion, Soviet authorities were desperate to contain the radioactive ruins. A concrete-and-steel sarcophagus was hastily built. Unfortunately, owing to its rushed construction, it didn't properly seal the reactor, prompting fears that the casing would collapse and allow more radioactive materials to escape into the atmosphere. A second clean-up was required, along with a new, permanent sarcophagus, funded by a €2.1 billion 'New Safe Confinement' project launched in 1997. French construction consortium Novarska set to work in 2011 with a labour force 10,000 strong to build a massive arch – intended to last for 100 years – that will encase toxic materials and allow other teams to dismantle the earlier sarcophagus by 2023.

The total cost of Chernobyl is now estimated at about $235 billion and scientists estimate that the area could remain uninhabitable for humans for 3,000 years. Chernobyl stands as a cautionary tale of the dangers of uncontained nuclear power, a human disaster that shattered lives and families, and an environmental catastrophe.

The 1986 nuclear disaster at Chernobyl has left an indelible mark on history. Unfortunately, it was not to be the last of its kind.

CHAPTER 21

EASTERN ENIGMAS: CHINA, INDIA, PAKISTAN – AND ISRAEL

'The Nuclear NPT [Non-Proliferation Treaty] disintegrates before our very eyes . . . the current non-proliferation regime is fundamentally fracturing. The consequences of the collapse of this regime for Australia are acute . . . including the outbreak of regional nuclear arms races in South Asia, North East Asia and possibly even South East Asia.'
– **KEVIN RUDD, AUSTRALIA'S THEN SHADOW MINISTER FOR FOREIGN AFFAIRS, SYDNEY INSTITUTE, 2006**

We tend to forget that there are a number of other vulnerable nuclear areas that, all too often, escape our scrutiny. Not every nuclear accident has occurred in the so-called 'developed world'. There are ever more nuclear powers out there, each with their own, little-known histories of nuclear blunders, mistakes and accidents.

Chinese efforts to develop a bomb started early. Following the Communist Revolution of 1949 and Chiang Kai Shek's flight to Formosa (Taiwan), Mao Zedong realised that he dare not attack the Chinese nationalists for fear of what he termed,

'Washington's atomic blackmail'. Mao was further checked in July 1950, when US President Harry Truman ordered nuclear bombers to the Pacific, with the intention of deterring Chinese intervention in the Korean War.

In 1954, as the Korean peace talks stalled, the commander of the US Strategic Air Command, General Curtis LeMay, was in typically bellicose form, stating, 'We have no suitable strategic air targets in Korea. However, I would drop a few bombs in proper places like China, Manchuria and southeastern Russia. In those "poker games", such as Korea and Indo-China, we ... have never raised the ante – we have always just called the bet. We ought to try raising sometime.' This naked threat spurred the People's Republic on to build its own bomb.

In October 1957, with Soviet support and technical data from which Beijing could manufacture its own nuclear weapon, using weapons-grade uranium from a new plant in Lanzhou, China began to build a primitive bomb. Beijing tested it on 16 October 1964 and became the fifth nuclear power. It still lacked any launch capability. Even that was solved by the still-friendly Soviets, who gifted their neighbour two primitive R-2 missiles (an improved version of the German V-2 rocket) as models from which to work. In 1966, China launched a Dong Feng-2 (DF-2) medium-range ballistic missile with a 12-kiloton nuclear warhead from the Shuangchengzi Missile Test Site, which struck its target on the Lop Nur Test Site.

Since then, China has pursued what appears to be a pragmatic and rational approach to a nuclear deterrent based on the concept of no-first-use. Beijing has consistently insisted that the country's nuclear weapons are intended only as a minimum deterrent against nuclear attacks. Although the exact size of China's nuclear stockpile remains a state secret, CIA estimates in a 2011 report indicated that the country had only produced 200–300 nuclear warheads, despite rumours of secret stockpiles.

Today, China is building new missile launch sites and underground storage facilities far inland. These launch sites appear to be intended primarily to guard against Russia and India, as there is no evidence from satellite intelligence of long-range missiles being deployed to these new locations. This makes sense when applied to China's wider nuclear policy, which is to build up a 'nuclear Triad', a mixed force consisting of land-based ICBMs, submarine-launched ballistic missiles (SLBMs) and strategic bombers capable of fast retaliation. The nuclear Triad weapon mix is a potent deterrent.

The 2010 China Defence White Paper very clearly spelled out Beijing's no-first-use policy. It states China's 'unequivocal commitment that under no circumstances will it use or threaten to use nuclear weapons against non-nuclear-weapon states or nuclear-weapon-free zones'. Moreover, the director of the Chinese Academy of Military Science subsequently reiterated that there is 'no sign that China is going to change a policy it has wisely adopted and persistently upheld for half a century,' and China re-affirmed its no-first-use policy in its most recent 2020 Defence White Paper.

However, where once the People's Republic was a trailblazing pioneer for civilian nuclear power, its wider atomic ardour seems to have cooled somewhat. Although China lacks serious domestic energy supplies, only 4 per cent of China's electricity comes from nuclear energy. Coal-fired power stations are still preferred, carbon emissions and global warming notwithstanding. This growing wariness of matters nuclear seems to have been inspired by a series of nuclear mishaps. In the wake of Japan's 2011 nuclear crisis at Fukushima, both China and Taiwan moved swiftly to cooperate on nuclear safety issues, including establishing a formal nuclear safety agreement and an official contact mechanism between the two sides, which is used as a nuclear civilian hotline to facilitate information and emergency responses in case of an accident.

The Fukushima disaster, the worst nuclear accident in twenty-five years, also forced a serious re-think of China's wider nuclear ambitions. It sent authorities scrambling to reassure its own people that they were not at risk of a similar catastrophe and sparked an immediate moratorium on new power plants.

China is still the world's largest consumer of energy, thanks mainly to its burgeoning industrial capacity. This is only going to increase over the next decades with households expected to use nearly twice as much energy as they do presently by 2040, according to the International Energy Agency. But long-term Chinese plans for more nuclear power plants seem to be on hold at the time of writing.

It all comes down to the profound shock of Japan's Fukushima disaster. What was described as 'a sudden aversion to nuclear energy' caused the Central Committee's State Council to suspend approval for dozens of nuclear power projects and institute an immediate safety inspection of all existing nuclear facilities. New regulations were passed, including the '2020 Vision for Nuclear Safety and Radioactive Pollution Prevention'. According to Mark Hibbs, an analyst with the Carnegie Endowment and co-author of the 2012 report *Why Fukushima Was Preventable*, 'Some semi-official projections that China might have more than 400 nuclear plants by 2050 have been cut in half.'

Beijing's nuclear energy policy masks some other challenges facing China's governing party and their atomic plans. Ordinary Chinese people are now a lot less enthusiastic about nuclear power than their leaders imagined. There is also growing evidence of protests against proposed nuclear plants. 'Not In My Back Yard' (NIMBY) activity is not just confined to the West. A government-supported survey by the Chinese Academy of Engineering in August 2017 found that 'only 40 per cent of the public now supports the development of nuclear power in China'. It went on, 'The Fukushima accident has had the consequence that the public has become more sensitive to the possible development of nuclear energy projects, and is opposing such projects, especially near their homes.' Evidence of this growing resistance to nuclear power was revealed in 2017, when plans to build a nuclear-waste processing plant in the eastern province of Jiangsu resulted in violent protests from local NIMBYs and the project was eventually scrapped.

Given the potential ramifications of a nuclear disaster in the world's most populated country, to say nothing of economic or environmental fallout, this conclusion seems both credible and sensible. In the densely populated parts of eastern and southern China where many of the country's nuclear reactors are located, any nuclear exclusion zone could affect huge numbers of people, as the prevailing winds could carry radioactive fallout into the rest of China, Korea or even Japan. The consequences of a nuclear accident hardly bear thinking about.

The irony is that China's nuclear industry has a generally good safety record and can point to the fact that in four decades of operating nuclear plants, it has never experienced, or reported, a major accident. Despite this, many Chinese are now openly admitting that the danger of accidents poses a serious threat to the future of nuclear power.

The wider conclusion must be that nuclear power is no longer the attractive, clean energy solution for the PRC that it once was. Miles Pomper, a Washington-based expert in nuclear energy at the James Martin Center for Nonproliferation Studies, sums it up: 'For a long time, China was basically subsidising the [nuclear] industry, and now they're trying to put it on a market footing. When you do that, oftentimes it doesn't meet the market test, especially competing with wind and other kinds of power.'

According to energy journalist Peter Fairley, 'If China's nuclear ambitions wind down, it may be the nail in the coffin for the technology's viability elsewhere.'

On China's western border, another nuclear state, India, appears to have different ideas. India had nuclear ambitions from the start. As early as 1946, the country's first prime minister, Jawaharlal Nehru, declared, 'I have no doubt India will develop her scientific researches and I hope Indian scientists will use the atomic force for constructive purposes. But if India is threatened, she will inevitably try to defend herself by all means at her disposal.'

The brief 1962 Himalayan war with China spurred India to develop its own bomb. This was first tested in 1974 using Canadian-supplied plutonium. India performed more nuclear tests in 1998 and a worried USA even slapped temporary sanctions on Delhi.

An Indian Strategic Nuclear Command (SNC) was formally established in 2003, although India has a public no-first-use policy and a nuclear doctrine based on what it terms 'credible minimum deterrence'. The SNC reports to the Cabinet Committee on Security, the only body authorised to order a nuclear strike. However, nuclear ambiguity crept in when in 2019 the country's defence minister hinted that in future India's 'no-first-use' policy might change 'depending upon the circumstances'. To back this, Delhi has developed an embryonic Triad of strike bombers, land-based missiles and nuclear submarines to deter any aggressor.

While India's military nuclear aspirations seem clear-cut and effective, their civilian nuclear programme has not been either as successful or as safe. Delhi's record on nuclear power is far from reassuring. In 1987, the Fast Breeder Test Reactor at Kalpakkam suffered a serious refuelling accident that ruptured the reactor core, resulting in a two-year shutdown. In 1989, operators at the Tarapur Atomic Power Station found that the reactor had been leaking radioactive iodine at more than seven hundred times normal levels. Repairs took more than a year. Then, in 1993, the Narora Atomic Power Station in Uttar Pradesh nearly had a complete meltdown as a fire at two of its steam turbines damaged the heavy-water reactor and stopped the vital cooling.

In 2003, India suffered by far its worst nuclear accident. Six workers at India's Kalpakkam Reprocessing Plant (KARP) were exposed to what was described as a 'severe' dose of radiation. Mr B Bhattacharjee, director of BARC (the Bhabha Atomic Research Centre), admitted, 'This is the worst accident in radiation exposure in the history of nuclear India.'

The incident led to the indefinite shutting down of the plant, and raised serious questions over the safety of the production of potential weapons-grade plutonium at KARP, and also the safety of workers and human habitations around Kalpakkam, seventy kilometres from Chennai. Mr Bhattacharjee admitted that 'some [of the workers] were not even wearing badges and one of them was a woman'. (The badges he referred to are used to monitor the levels of radiation to which a worker in such environments has been exposed.) He refused to specify precise figures for the amount of radioactivity to which people had been exposed, stating that they were higher than the maximum allowed annual dose but 'lower than the maximum allowed lifetime dose'. It emerged that two previous incidents at the KARP facility in 2001 and 2002 had also exposed workers to levels of radiation that had been kept concealed. Not unreasonably, the plant workers voted with their feet and went on strike. It later emerged that two workers had received a dose of 420 rads (any more than 100 rads is a dangerous dose) and others reportedly received overexposures in the range of 280 rads, all grossly exceeding the permissible radiation dosage limits, both annual and cumulative. India's attitude still remains astonishingly casual. Director of the Kalpakkam facility, Mr S Basu, was dismissive of the dangers. 'If you accept atomic energy, you cannot shy away from radioactivity,' he said. 'If you work in a radioactive area, you are bound to get some radiation exposure.'

The potential danger persists. In late 2019, Delhi admitted that that there had been a serious cyber attack on the Kudankulam Nuclear Power Plant (KKNPP) in Tamil Nadu, the biggest such facility in India. Officials denied that there had been any damage or problems but the *Washington Post* noted that hacking 'can be used to facilitate sabotage, theft of nuclear materials, or – in the worst-case scenario – a reactor meltdown'. In a densely populated country such as India, any radiation release from a nuclear facility would be a major disaster. India's safety record and apparently casual attitude to nuclear safety do not bode well for the future. On the evidence so far, its nuclear programme looks like an accident waiting to happen.

India's regional adversary, Pakistan, came to the nuclear club late. Following the loss of East Pakistan in the Bangladesh Liberation War of 1971, Pakistan's Prime Minister Bhutto began 1972 with a gathering of senior scientists and engineers. During this 'Multan meeting', Bhutto encouraged the scientists 'to build an atomic bomb in three years for national survival' and with a commitment to having materials for a nuclear bomb ready by the end of 1976.

The progress of Pakistan's nuclear weapons programme was hastened when India developed its 'Smiling Buddha' nuclear test in 1974, the first time that a nation not permanently part of the United Nations Security Council had carried out a confirmed nuclear test. A few weeks after India's second nuclear test, in May 1998, Pakistan became the seventh country to successfully develop nuclear weapons, choosing the north-west Baluchistan desert as a location for detonating five nuclear devices, containing fissile material from the Khusbab Nuclear Complex refinery for weapons-grade plutonium.

India's rival China was a keen backer of the Pakistani nuclear programme from the start. Bhutto visited Beijing to cement strong ties between Chinese and Pakistan scientists. Chinese aid was reciprocated when, in 1986, scientists and engineers from the Pakistan Army Engineers helped Beijing develop new weapons-grade centrifuge technology for Chinese nuclear weapons and even constructed an enrichment plant at Hanzhong.

Pakistan has five main reactors for civilian energy production, and older plants at Kanupp and Chashma, but they only contribute about 2 per cent of the country's energy demands, and while a second nuclear reactor was established at Chashma in 2017, they are expensive to construct and maintain, at around $1 billion each. Although the budget for the development of nuclear energy in Pakistan has been cut by almost half to Rs15.08 billion ($142 million) in recent fiscal years, almost half is allocated to pay back the interest on construction project building loans.

These reactors have been costly but ineffective. The Kanupp reactor, for instance, constantly under repair, is only operating at 30 per cent capacity. 'It generates enough electricity,' believes physics professor A H Nayyar, 'to power just 3 per cent of Karachi.' Nuclear physicist Dr Pervez Hoodhboy is equally critical of such reactors, describing them as 'nothing more than toys'.

But building new reactors still continues even after the Chinese-designed reactor at the Chashma Nuclear Power Plant was connected to the grid to become the fifth nuclear reactor in Pakistan. To make matters worse, Pakistani reactors have poor safety records. Kanupp is located next to the Arabian Sea and despite an official denial at the time, we now know that some thirty years ago radioactive cooling water escaped from it, by accident. In 2011 the Kanupp Karachi power plant experienced a leakage of heavy water from a feeder pipe to the reactor, and had to declare an emergency that lasted seven hours.

Notwithstanding Islamabad's flaky record on nuclear power, the government bureaucrats of Pakistan's Atomic Energy Commission ensure that the subject remains shrouded in secrecy. There is little public debate about the country's nuclear safety record, even in the wake of Fukushima.

But the original Pakistani plants are now rather old. Kanupp, commercially established in 1972, whose lifespan was only intended to be thirty years, has now been granted a further ten years of operation by the country's Ministry of Defence. To quote Dr Hoodhboy: 'The operators working there privately say that this reactor has gone beyond its life and they are afraid that something could go wrong. The structure has been weakened by decades of radiation.'

The IAEA safety monitoring is perfunctory, according to Dr Hoodhboy, as they 'do not have the capacity to look at everything important *inside* a reactor. Spent fuel is stored onsite and should an accident occur, the devastation would be very great.' He likens the site to a ticking bomb. 'Should an accident occur, the coastal winds could blow the radioactive plume over Karachi, which has grown to a population of nearly 18 million in the past forty years and there are homes very close to the reactor now. There is the absence of a safety culture. Then there is the incapability of the authorities to deal with anything of this magnitude,' says Dr Hoodhboy.

The reactors in Chashma are no safer. Reactors Chasnupp I and II are on the banks of the Indus River, Pakistan's lifeline and a major source of fresh water for irrigation and domestic use by millions. The whole site is built on a known seismic fault line with a long record of earthquakes – just like Fukushima. The uncomfortable truth is that nuclear power on the subcontinent is not a subject to inspire confidence.

Further west is the great nuclear enigma, Israel. The tiny Jewish state, clinging to the edge of the Levant, maintains a beleaguered existence in a sea of hostile Arabs and Muslims sworn to its destruction.

Israel's nuclear story started badly. By 1963, the country had entered into a major feud with US President John F Kennedy. Jewish supporters and agents owned or controlled large chunks of the American media, big banks and corporations. These 'friends of Israel' monitored virtually all aspects of US life at both national and regional level. The respected senator J William Fulbright summed it up in 1973:

Israel controls the US Senate. The Senate is subservient, in my opinion much too much. We should be more concerned about the United States' interest rather than doing the bidding of Israel ... The great majority of the Senate – somewhere around 80 per cent – are completely in support of Israel, anything Israel wants. This has been demonstrated time and again and has made [foreign policy] difficult for our government.

In 1957, David Ben-Gurion, Israel's first prime minister, had ordered a disguised nuclear facility to be built at Dimona in the Negev Desert. Construction work at the site started in 1958, with Chinese help. Although Tel Aviv had tried to keep the Dimona nuclear plant a secret, the CIA was well aware of the clandestine nuclear programme. When, in 1961, President Kennedy discovered that Israel was secretly building a nuclear bomb, he was furious. Stopping the spread of nuclear weapons in the Middle East was more important than blindly supporting Israel, as far as Kennedy was concerned. He wanted to inspect Dimona.

Israel's prime minister did not agree. The ageing Ben-Gurion had spent his life working to create a Jewish homeland in Palestine. To him, weapons of mass destruction were essential for Israel's survival. Israel's 'secret' nuclear plans started what was to become one of the most poisonous relationships ever recorded between two supposedly friendly heads of state, one that grew into open enmity through layers of deceit and backstairs manoeuvring. The prime minister told falsehoods both to his own countrymen and to JFK. In 1961, Ben-Gurion met Kennedy in New York. At that meeting, both men lied to each other.

That face-to-face meeting sowed the seeds of a deep mutual dislike. As his anger grew with the Israeli dishonesty and prevarications over the existence and purpose of Israel's Dimona nuclear facility, so Kennedy flexed his muscles. He suspected that an American company called the Nuclear Materials and Equipment Corporation (NUMEC) was transferring US fissile material. NUMEC was owned by an American Jew named Zalman Shapiro, who clearly had divided loyalties. After Kennedy's death, the Atomic Energy Commission found that Mr Shapiro had managed to 'lose' some 25–100 kilograms of enriched uranium. According to the entry for Shapiro on Wikipedia: 'No charges were ever filed, and one report concluded that there was "no substantive evidence to indicate that a diversion occurred". Shapiro denied any wrongdoing, and said that such discrepancies are "not unusual" and that losses could be explained as normal to the complex processing.'

In 1963, Kennedy's demands to verify the true purpose of Dimona grew terser and more threatening. Ben-Gurion tried mobilising his vociferous pro-Israel lobby in the States to thwart the President, but to no avail. Kennedy was adamant. He wanted his nuclear inspectors to check out Dimona – properly. Just why was an 'agricultural research establishment' being run by the Ministry of Defence? And why was it so tightly guarded?

Ben-Gurion finally caved in. However, he had one trick up his sleeve to deceive the ever-persistent Americans. He ordered a dummy plant built nearby, and – in a letter dated 12 May 1963, insisted 'Mr President, my people have the right to exist ... and this existence is now in danger.' JFK was furious. On 17 May 1963, he blasted off a 'tough and extremely threatening' letter, pointing out bluntly that if Ben-Gurion did not do what he required, then '[the USA's] commitment to and support of Israel would be seriously jeopardised.'

For Ben-Gurion, this was the last straw. He envisaged his historic legacy as Jewish patriarch and the reputation as the founder of Israel slipping away. At seventy-seven years of age, he suddenly saw his life's work – building the Jewish State of Israel in his own image – threatened by some callow American who knew nothing about Israel and the harsh realities of Jewish survival in the Middle East. As far as he was concerned, the US President had now joined with Israel's many foes and was actually threatening the very existence of the country he had built. JFK was undeterred, writing another demanding letter on 15 June. It went undelivered – Ben-Gurion dramatically resigned the next day – but the letter clearly served as the model for another that Kennedy wrote to his successor, Levi Eshkol, who received it on 5 July. It was regarded by all parties as effectively an ultimatum:

> Dear Mr Prime Minister...
> As I wrote Mr Ben-Gurion, this Government's commitment to and support of Israel could be seriously jeopardised if it should be thought that we were unable to obtain reliable information on a subject as vital to peace as the question of Israel's effort in the nuclear field...
>
> It would be essential... that our scientists have access to all areas of the Dimona site and to any related part of the complex, such as fuel fabrication facilities or plutonium separation plant, and that sufficient time to be allotted for a thorough examination.
>
> Knowing that you fully appreciate the truly vital significance of this matter to the future well-being of Israel, to the United States, and internationally, I am sure our carefully considered request will have your most sympathetic attention.
>
> Sincerely,
> John F Kennedy

The threat was clear. If Israel did not allow American nuclear inspectors into the secret site at Dimona, the White House would cut off US aid. Without American money and surrounded by Arabs sworn to its destruction, Israel would be unlikely to survive.

JFK's assassination on 22 November 1963 solved Israel's problem at a stroke. His successor, Lyndon B Johnson, commonly known as 'LBJ', was the most pro-Israeli president of all and turned the money taps on full. By 1966, US military aid to Israel had tripled to an all-time high and Israeli nuclear research accelerated unhindered. We now know that – concerned that he would lose the Jewish vote – President Johnson smothered the 'Gilpatric Report', which was published in January 1965 and called for action against several states with nuclear weapon ambitions, including Israel. He went further still in burying the report in June that year when Senator Robert F Kennedy publicly backed many of its recommendations, invoking the name of his assassinated brother John.

Confirmation that Israel was working on a nuclear bomb came on 22 September 1979, when the US surveillance satellite Vela 6911 recorded an unusual double flash as it orbited Earth above the Indian Ocean. It was the unmistakable signature of a nuclear explosion – something US satellites had detected on dozens of previous occasions in the wake of similar tests. There was only one prime suspect: Israel, at the time deeply in cahoots with the South African government. On the fortieth anniversary of the event, a US foreign policy team analysed the declassified documents on 'the Vela incident' to explain that mysterious flash four decades before. They decided that it was an Israeli nuclear test, proof that Israel's bomb had worked.

The result of Washington turning a blind eye to Israel's nuclear ambitions is that today Israel is undoubtedly a nuclear power with a serious arsenal. Along with India and Pakistan, Tel Aviv has never signed the international Non-Proliferation Treaty (NPT) and possesses between eighty and four hundred nuclear warheads. These are divided among a land, sea and air Triad weapon mix, giving a strike force with the ability to deliver nuclear weapons by aircraft; as submarine-launched cruise missiles; and via the 'Jericho' series of intermediate to intercontinental range ballistic missiles. These ICBMs are not just regional. At one point during the Cold War the Americans discovered, somewhat to their alarm, that Israel, on discovering that Moscow had four ICBMs targeting Israeli cities, had returned the compliment as a deterrent to Moscow meddling in Israel's Middle East wars.

Although Israel remains paranoiac about her military nuclear capability, even drugging and kidnapping a runaway nuclear technician, Mordechai Vanunu, then jailing him for eighteen years to stop him talking, Tel Aviv has not ignored civil nuclear power. In 2011, after the nuclear emergency at Japan's Fukushima Daiichi Nuclear Power Plant, Israel's then prime minister, Benjamin Netanyahu, stated, 'I don't think we're going to pursue civil nuclear energy in the coming years.' The truth is that Israel has abundant offshore gas available and the Dimona complex's prime task is really to produce the fissile material needed for nuclear weapons.

Worryingly, Dimona may be the world's oldest reactor. Dr Uzi Even, one of the scientists who worked at Dimona from the start, warned in April 2016 that the facility, which has been operational since the 1960s, should be shut for safety reasons. After a report appeared, warning that ultrasound tests had revealed 1,537 defects and flaws in the aluminium core, he told Israel Radio, 'It doesn't come as a surprise. I've been warning about this possibility for many years.'

The nuclear physicist also criticised the limited monitoring at the nuclear site. As Israel is not a signatory to the NPT, it is not subject to scrutiny by the International Atomic Energy Agency, although it claims to operate in accordance within the IAEA's safety regulations. The problem, according to Even, is that Israel cannot manufacture the parts needed to repair the site. It also lacks the capabilities to build a new reactor, and Even feels that unless Israel changes its approach and signs the NPF, 'There is no one in the world who will agree to sell us such a reactor . . . It had a very specific goal when it was founded in the Sixties. But fifty years is a long time, and a lot has changed,' he declared.

On the surface, Dimona has a good safety record, but Israeli secretiveness makes it hard to judge. For example, only in 1994 did full information emerge about a laboratory accident there in December 1966, which caused one fatality and at least one other person to be injured.

There had been denials from Israeli military censors that any explosion had occurred or that radioactive material was removed during the two-month clean-up following the accident. But eventually it was Environment Minister Yossi Sarid who revealed the truth: 'There was a blow-up within the laboratory . . . There was a certain degree of radiation [inside the laboratory], but outside there was no radiation. It was cleaned up.' Furthermore, Israeli television broadcast an interview with a site employee who reported that the accident had caused a fatality. The identity of the deceased was discovered in newspapers, by the journalist Alex Doron, as being twenty-two-year-old lab technician Abraham Gofer.

It was not the only accident. Radioactive material has leaked from the Dimona nuclear reactor several times since it was built, according to the news site Arab48. This was revealed during a case against the reactor filed by a former worker who was battling cancer. Faridi Taweel claimed that his cancer was caused by radioactive leakage from the reactor. Otherwise, Dimona keeps its secrets.

Israeli has good cause to be enigmatic about the nuclear threat that it faces. The tiny country is surrounded by foes and now they are well-armed and sitting, thanks to Lebanon's Hizbollah and the Iranian Revolutionary Guard, close to home on Israel's very borders. Ever since the surgical air strike to neutralise dictator Saddam Hussein's nuclear reactor and planned atomic bomb facility at Osirak in Iraq in June 1981, Israel's policy remains unequivocal: if an Arab leader calls for the destruction of Israel, Israel will not allow them to have nuclear weapons.

Iran has openly called for the 'Jewish State' to be driven into the sea. Tehran's hostile words are easy to ignore, but Tehran tooling up for a nuclear strike is different. Iran's twenty-first-century drive to acquire a bomb sent shock waves through Israel and has ensured a struggle to the death with nuclear weapons as the key.

Iraq's ambitions to go nuclear began to take off as far back as 1967, when the Tehran Nuclear Research Center acquired a small American pilot reactor. By the mid-1970s, Iran had begun developing a civilian nuclear power programme. All that stopped in 1979 with the Iranian Revolution and the fall of Shah Mohammad Reza Pahlavi.

In 1984, China stepped in to fill the gap, supplying a nuclear research centre in Isfahan. By 1998, an increasingly worried Washington was expressing concerns that Iran's nuclear energy programme could lead to the development of nuclear weapons. Since 2003, worldwide concern over Iran's nuclear programme has increased, although Iran's Supreme Leader, Ayatollah Ali Khamenei, repeatedly denied Iran was building a bomb, even claiming that weapons of mass destruction are forbidden under Islamic law. This egregious claim was undermined by the CIA discovery in 2005 that Iran had restarted uranium conversion, a step on the way to enrichment.

Iran has always insisted its nuclear programme is peaceful, but suspicions that it was being used as a cover to develop a nuclear bomb prompted the UN Security Council, USA and EU to impose sanctions from 2010. That was the year that the Israelis and the US launched the crippling Stuxnet computer worm, specifically designed to wreck Iran's nuclear centrifuges, busy refining weapons-grade plutonium. The undeniable cyber-attack put any Iranian nuclear programme back by three to five years. By 2014, a report based on IAEA data warned that Iran would soon be able to produce weapons-grade uranium. In 2015, Iran reached a deal with the USA, UK, France, China, Russia and Germany to limit Tehran's nuclear activities, in return for sanctions relief. The deal restricted Iran's enrichment of uranium. Tehran was also required to allow international inspections and to redesign a heavy-water reactor under construction, because spent fuel would contain weapons-quality plutonium.

Iran was quickly suspected of cheating and US President Donald Trump abandoned the deal in 2018 to reinstate heavy US sanctions. Along with his suspicions about North Korea's secretive bomb-making efforts, Trump wanted to squeeze Iran to curb any ballistic-missile programme and stop its involvement in regional conflicts. Tehran refused: the impact of US sanctions inside Iran was devastating. Oil sales collapsed, the value of the Iranian rial crashed, and inflation soared as US economic pressure took effect, causing widespread social unrest.

As ISIS's Islamic State collapsed back into a jihadi terrorist organisation in 2019 and Tehran's troops moved into Syria to fill the gap and further fuel Syria's long-running and bloody civil war, all thoughts of any peaceful US-Iranian deal collapsed. By 2020, a state of undeclared hybrid warfare rocked the Middle East. Tensions escalated with tit-for-tat drone strikes and Iranian missile attacks on US troops.

In this dangerous muddle, Israel became increasingly tense and rightfully watchful, warning that any attempt by Tehran to attack Israeli cities would be met by savage retaliation. More dangerous still, Jerusalem warned that any attempt by Tehran to acquire a nuclear-strike capability was unacceptable and would be met with 'an appropriate response'. The nuclear chips were now down in a volatile and explosive region. Pessimists genuinely believe that it can only have one outcome, what the Israelis call the 'Samson Option': pulling the temple – and everything else – down around everyone's ears.

To the ayatollahs, that meant only one thing: by hook or by crook, Tehran needs to go nuclear. For the Shi'as, martyrdom can wait. But as of early 2020, who could tell what it meant for Kim Jong-un and the enigmatic North Koreans quietly assembling a nuclear arsenal?

CHAPTER 22

JAPAN IN THE DOCK

'Japan knows the horror of war and has suffered as no other nation under the cloud of nuclear disaster. Certainly Japan can stand strong for a world of peace.'
– MARTIN LUTHER KING, JR, PAPERS PROJECT

Japan's stated position on nuclear weapons is as opaque as that of Iran and Israel. As the victim of the only two nuclear attacks in history, you might think it reasonable to assume that the country would be wary, if not downright nervous, about having anything to do with matters nuclear.

You could not be more wrong.

The Japanese record of nuclear accidents since 1945 is surprising because it is so appalling – one of the worst on the planet. It is difficult, given the lack of accurate statistics, to estimate Japan's total casualties as a result of the two atom bombs that fell on Hiroshima and Nagasaki that year. The destruction of hospitals, civil amenities, such as fire, police departments, and government agencies added to the state of utter confusion immediately following the explosions, plus the great fires that raged in each city, destroying many bodies, make it hugely problematic to work out how people many died.

The best estimates, according to the Atomic Archive, are about 200,000 total dead and countless short- and long-term casualties. Yet, despite this traumatic experience, since World War II Japan has relied heavily on nuclear power and adopted a very ambivalent attitude to nuclear weapons. Going nuclear for electricity is understandable. Japan has no indigenous fuel supplies and must rely on imports or hydroelectric generation for energy. Before the Fukushima accident, Japan was the third-largest nuclear power generator in the world behind the United States and France. However, the country had temporarily lost all of its operating nuclear generation capacity by 2013, as its facilities were removed from service because of the earthquake damage that brought about the infamous disaster at Fukushima. The composition of fuel used for power generation shifted heavily to fossil fuels, particularly liquefied natural gas (LNG), which became the primary substitute for nuclear power. But nuclear *weapons*?

Immediately after World War II, Japan was understandably hostile to all forms of nuclear power, be they military or civil. Its post-war constitution, devised with US help, firmly forbade the establishment of anything other than military forces for 'self-defence'. Things began to change, and Japanese attitudes to shift, in the years after China conducted their first nuclear test in 1964. On meeting President Lyndon Johnson in early 1965, Japanese Prime Minister Eisaku Satō expressed concern that if it was true that the Chinese Communists had nuclear weapons, Japan might then have to follow suit. Johnson and Washington were startled by this announcement, not least because Satō insisted that younger members of the public could be persuaded or 'educated' into supporting the plan. For Japan, it appeared that a psychological nuclear-weapons Rubicon was about to be crossed.

But the idea that Japan might become a nuclear power had been planted. In response, Washington aimed to grant Japan membership of the NPT, while, in December 1967, Tokyo announced the 'Three Non-Nuclear Principles', to dismiss production, possession or introduction of nuclear weapons as reassurance to the Japanese people. Though not officially passed into law, these principles have remained to this day as Japan's nuclear policy. All the same, at the time Satō persisted with further discussions, arguing that Japan could back a nuclear option as part of a policy of self-defence, and even the option of tactical nuclear weapons for defensive purposes only, for which a White Paper was published (which insisted that such an introduction would not violate the country's constitution).

Subsequently, following hostile foreign and domestic reaction, Tokyo backed away, declaring that its policy was not to acquire nuclear weapons 'at present'. These military policy musings were light years away from Japan's staunch post-war pacifism and opposition to all things nuclear.

No such constraint prevented the acquisition and use of nuclear power for generating electricity, however. The first Japanese nuclear power station at Tōkai near Tokyo was built in the early 1960s to the British Magnox design, and was up and running by 1966.

In 1968, in a speech to the Diet, Prime Minister Satō spelled out Japanese nuclear policy in explicit detail in his declaration of the 'Four Nuclear Policies' (also known as the 'Four-Pillars Nuclear Policy'):

- To promote the peaceful use of nuclear energy.
- To work towards global nuclear disarmament.
- To rely and depend on US extended deterrence, based on the 1960 US-Japan Security Treaty.
- To support the 'Three Non-Nuclear Principles' if and

when Japanese national security is guaranteed by the previous three Nuclear Policies.

Sharp-eyed diplomats quickly spotted that this left the nuclear weapons option available if American assurance was ever removed or seemed unreliable. Japan quietly retained the option of going nuclear *in extremis*. Tokyo finally ratified the NPT in 1976, but only after the USA promised 'not to interfere with Japan's pursuit of independent reprocessing capabilities in its civilian nuclear power programme'. This was sophistry. The link between civilian reactors and nuclear weapons is a very fine one, as Iran has demonstrated many times in its quest for a bomb.

In all fairness, Japan had good reason to concentrate on its pursuit of nuclear power. The country lacks natural resources and remains largely dependent on imported fossil fuels such as oil, coal and liquefied natural gas (LNG) to satisfy its energy demands. Ratios of dependence on imported fossil fuels in 2018 were high: 99.7 per cent for oil; 97.5 per cent for LNG; and 99.3 per cent for coal.

Nuclear therefore remains an important source of power. By the end of 2017, Japan had forty-two operable nuclear reactors, 22 per cent down from fifty-four in 2010. Ironically, the Fukushima Daiichi disaster merely emphasised the importance of nuclear power for Japan. More than 10 gigawatts of nuclear capacity at the Fukushima, Onagawa, and Tōkai facilities ceased operations immediately following the earthquake and tsunami. Other reactors were permanently damaged from emergency seawater-pumping efforts and several have taken years to return to service. Public anti-nuclear sentiment shifted markedly after 2011 and there were widespread public protests and calls for nuclear power to be abandoned completely. The government officially decommissioned all six reactors at the Fukushima Daiichi nuclear plant. By May 2012, for a short period, Japan did not have any nuclear generation for the first time in more than four decades.

Japan's 2014 new 'Strategic Energy Plan' tried to get policy back on course. It stated that nuclear remained an important source of power, but that dependence on nuclear generation 'will be offset as much as possible by alternative energy supplies'. By 2017, work had begun on restarting twenty-five reactors and building the new nuclear plant at Ōma, representing more than half of Japan's remaining operable capacity. The timeline for restarting many of these reactors was slow because of tighter regulations and the government's need to restore public confidence in nuclear power, but Japan was clearly intent on continuing building up its civil nuclear economy. The national economy needed its own power.

The balance between populist sentiment and the continuation of reliable and affordable electricity supplies was slowly being worked out politically. The government's aim remained for nuclear power to provide up to 22 per cent of Japan's electricity by 2030. At the time of the tsunami in 2011, nuclear energy accounted for almost 30 per cent of the country's total electricity production. At the time, there were plans to increase this to 50 per cent by 2030.

Tokyo has since made a determined effort to tighten up on nuclear generation and safety. In October 2012, the new Nuclear Regulation Authority (NRA), which took over from the discredited Nuclear and Industrial Safety Agency (NISA) and the Nuclear Safety Commission (NSC), announced that henceforth all nuclear-power-plant restart reviews would comprise both a safety assessment by the NRA and the briefing of affected local governments by the operators.

The NRA needed to show a firm grip because, Fukushima aside, historically Japan's safety record has been very bad. There have been far too many examples of accidents, human error, lax safety controls and regulatory failures over the years to begin to list in a single chapter. To take just a few examples from 1981's list of accidents gives some idea of the dreadful indictment of Japan's pre-Fukushima record on nuclear safety.

Over three days in January of that year, according to the National Resources and Energy Agency, forty-five workers at Tsuruga were exposed to up to 92 millirems a day, while seventy-six workers were, over six days, exposed to 155 millirems. In March, a major accident at Tsuruga caused over 15,000 litres of highly radioactive water to leak, exposing fifty-six workers to radioactivity and ultimately polluting Urazoko Bay – an accident that took over a month to become public knowledge, and part of the reason for the closure of the Tsugura plant for six months from May. The plant was even found to have covered up a leak of thirteen tonnes of radioactive water in early 1975.

It wasn't just the Tsuruga plant. The plant at Hamaoka leaked radioactive waste water in July 1981, while the reprocessing plant at Tokaimura was closed down completely in September after unusual levels of plutonium were identified.

There were many other incidents in that single year, some of them relatively minor, but even so too many to list here. And accidents were still occurring years later. In September 1999, there was the serious case of the runaway reactor at Tokaimura, an accident rated Level 4 on the INES scale. This wasn't the first accident at Tokaimura, either. An earlier incident had occurred on 11 March 1997 in a nuclear reprocessing plant owned by Dōnen (or the Power Reactor and Nuclear Fuel Development Corporation, in full), when a small explosion occurred in a nuclear reprocessing plant. Windows were blown out, allowing radioactive smoke to escape into the atmosphere. On the following Thursday, workers used duct tape (!) to perform repairs on thirty broken windows and three doors that sustained damage and were leaking radioactivity. At least thirty-seven of the workers were exposed to elevated levels of radiation during the incident.

Japan's first serious nuclear accident and uncontrolled reaction was a result of using an excess of higher-enriched uranium in preparation of nuclear fuel for the reactor. As part of their PH241 Nuclear Studies programme, Stanford University researchers investigated the second 1999 accident in some detail. They made some alarming discoveries.

Against all the rules and common sense, workers broke safety regulations by mixing dangerously large amounts of treated uranium in open metal buckets, setting off a nuclear reaction. The event caused the death of two workers and forty employees were exposed to dangerous levels of radiation. Hundreds of residents living nearby were evacuated from their homes while the nuclear reaction continued, but were allowed home two days later. The Stanford report is damning in its detail: 'Workers at the JCO nuclear fuel processing plant poured a seventh batch of 18.8 per cent enriched uranyl nitrate solution, along with uranium-235, into a precipitation vessel. The goal of this process is to convert the isotopically enriched uranium hexafluoride into uranium dioxide fuel [this] caused a self-sustaining chain reaction, otherwise known as a criticality, that persisted for several hours.'

On 9 August 2004, another serious nuclear accident killed five people at the Mihama power plant. Seven people were also badly scalded when hot water and steam leaks from a broken pipe. Officials insisted that no radiation leaked from the plant, and there was no danger to the surrounding area.

This appalling record of Japanese nuclear safety – or lack of it – was brought into sharp focus on 11 March 2011 in Japan's worst nuclear nightmare: Fukushima.

CHAPTER 23

JAPAN'S NUCLEAR NIGHTMARE: FUKUSHIMA

'The Fukushima nuclear complex went on to become the worst man-made engineering disaster in all of human history, outside of war.'

– STEVEN MAGEE, *HEALTH FORENSICS* (2014)

When we say that we are 'standing on solid ground', we are not. In fact, we are standing on a very thin crust of cooled rock floating on a massive ball of boiling molten metal. This solid 'upper mantle' of the lithosphere, to give it its scientific name, averages about 80 kilometres thick. Below that is the Earth's core, 1,200 kilometres across, with an inner core of magma (molten rock) 640 kilometres deep, a self-sustaining white-hot ball of silicates, iron and nickel at a mind-boggling temperature of 8,000°C. By way of comparison, the temperature of the surface of the Sun is about 6,000°C.

This mantle of solid rock is effectively floating on top of a liquid core of molten magma, and from time to time the floating solid rock moves. Our cooled crust consists of several, separate continent-sized plates jostling for position as they float on top of the molten liquid underneath. Known as 'tectonic' plates, they rub up against each other far below our feet and occasionally collide with stupefying force. The reverberations from these collisions are called earthquakes. The joins between the tectonic plates are the weak cracks, or fault lines, in the Earth's crust.

Well-known fault lines lie on both sides of the Pacific Ocean tectonic plate; to the east is the notorious San Andreas fault, running through San Francisco and Los Angeles. On the other side of the Pacific 'Ring of Fire' is the Australia-Alaskan fault. Both are notorious for sudden volcanic eruptions of molten magma from far below – and earthquakes.

No one in their right mind would build on a known volcanic fault line. Unfortunately, a lot of people who should know better accept the hazard and do just that. California could one day slide into the sea and Japan could any day suffer the fate of Christchurch, New Zealand. Japan's long history of earthquakes does not inspire confidence in the buildings insurance business.

The Japanese know all about earthquakes. The country accounts for about 20 per cent of the world's earthquakes of a magnitude of 6 or more on the Richter scale. No earthquake has ever registered on the Richter scale above 9.1. About 1,500 earthquakes strike Japan every year. Minor tremors occur on a daily basis. A tremor occurs in the island nation at least every five minutes, and each year there are up to 2,000 quakes that can be felt by people.

The Great Kanto earthquake of 1 September 1923 (7.9 on the moment magnitude scale) killed more than 140,000 people in the Tokyo area. Seismologists have said another such quake could strike the city at any time. On 17 January 1995, a 7.3-magnitude earthquake devastated the port city of Kobe. It was the worst earthquake to hit Japan in fifty years, killing more than 6,400 people and causing an estimated $100 billion in damage. Given these facts, it is hard to understand why anyone would want to build a vulnerable nuclear power station within 800 kilometres of Japan.

Worst of all are the unseen earthquakes, offshore, far beneath the sea. They displace billions of tonnes of water and create massive tidal waves. On Friday, 11 March 2011 at 2:46 pm, a magnitude-9 earthquake took place deep under the Pacific, 372 kilometres northeast of Tokyo, at a depth of 25 kilometres. The earthquake actually moved Honshu (the main island of Japan) 2.5 metres, shifted the Earth on its axis by at least 15 centimetres and fractionally increased the planet's rotational speed. On the surface, the shockwaves caused a tsunami with 30-metre waves to roar outwards at 480 kilometres per hour. It was the largest recorded earthquake ever to hit Japan. Just under an hour after the earthquake, the first of several tsunami waves hit the coastline. They surged inland at up to 40 metres above sea level in Miyako city and travelled as far as 10 kilometres, to near Sendai, flooding an estimated area of approximately 560 square kilometres.

As the tsunami roared ashore it killed thousands of people. The Japanese National Police eventually confirmed 15,899 deaths, 6,157 hospitalised and 2,529 people missing. Four years later, in 2015, a reported 228,863 people were still homeless as a result of the tsunami.

That earthquake began Japan's worst nuclear disaster.

First commissioned in 1971, Fukushima was the first nuclear plant to be designed, constructed and run in conjunction with General Electric and the Tokyo Electric Power Company (TEPCO). There are actually two power stations at Fukushima: Daiichi ('first') and 'Daini' (second) plants, about thirteen kilometres apart on the coast, with Daini to the south.

The main plant was built on a cliff that was originally thirty metres above sea level. During construction however, TEPCO lowered the height of the cliff by ten metres so that the foundations of the reactors would be on solid bedrock in order to reduce any dangers from earthquakes. Architects' analysis of the tsunami risk when planning the site's construction determined that this lower elevation was well within known safety limits, because the sea wall would provide adequate protection for the most powerful tsunami that the designers assumed it would face.

Some Japanese scientists had been uneasy about the location of the plant at the time, although the designers had built in agreed safety measures based on forecasts of pre-2011 earthquakes. There were, however, other indicators of risk. Japanese historians and geologists pointed out that more than a thousand years before, in AD 869, a monster tsunami had struck the northern Honshu region. Judging from geological evidence, it had come ashore at Sendai on the coast near the Fukushima site, wiping out whole towns and depositing sand and shellfish five kilometres (three miles) inland. Officials responsible for the country's modern earthquake hazard assessments overruled these ancient worst-case estimates, instead opting for the 'most probable' risk. The plant went ahead and became fully operational in 1979, with six light-water reactors generating 4.7 gigawatts of power. The result was that Fukushima Daiichi became one of the world's fifteen largest nuclear power stations.

The disaster at Fukushima Daiichi remains a mystery because most of the damage to reactors occurred only during the forty-two minutes between the earthquake and the subsequent devastating tsunami. Surely if the earthquake affected the plant's structure and the safety of its nuclear fuel to such an extent, the safety of all other similar reactors across the country was, and is, similarly jeopardised. The tsunami only made a dangerous failure much worse. Was the earthquake or the tsunami to blame?

On 2 July 2011, investigative journalists Jake Adelstein and David McNeill reported in *The Atlantic* magazine that over the following months of equivocation and distortion of truth, one claim remained: 'The earthquake knocked out the plant's electric power, halting cooling to its reactors,' in the words of the government spokesman Yukio Edano, at a press conference in Tokyo on 15 March. 'The story,' wrote Adelstein and McNeill, 'which has been repeated again and again, boils down to this: "after the earthquake, the tsunami – a unique, unforeseeable [the Japanese word is *soteigai*] event – then washed out the plant's back-up generators, shutting down all cooling and starting the chain of events that would cause the world's first triple meltdown to occur."'

Along with fellow journalist Stephanie Nakajimab, Adelstein and McNeill dug deeper and discovered an alternative explanation of events – a very different story, which they exposed in *The Atlantic* magazine. From first-hand eyewitness testimony, they concluded that it was the earthquake that had started the nuclear breakdown and disaster. One maintenance engineer who was at the site on that day was adamant that he saw hissing, leaking pipes after the earthquake and nearly an hour before the tsunami struck:

> I personally saw pipes that came apart and I assume that there were many more that had been broken throughout the plant. There's no doubt that the earthquake did a lot of damage inside the plant. There were definitely leaking pipes, but we don't know which pipes – that has to be investigated. I also saw that part of the wall of the turbine building for Unit 1 had come away. That crack might have affected the reactor.

The reactor walls are quite fragile, he explained:

> If the walls are too rigid, they can crack under the slightest pressure from inside so they have to be breakable because if the pressure is kept inside and there is a build-up of pressure, it can damage the equipment inside the walls

so it needs to be allowed to escape. It's designed to give way during a crisis, if not it could be worse – that might be shocking to others, but to us it's common sense.

Another eyewitness, a technician who was also present at the time, gave this account of the quake:

It felt like the earthquake hit in two waves, the first impact was so intense you could see the building shaking, the pipes buckling, and within minutes, I saw pipes bursting. Some fell off the wall. Others ruptured. I was pretty sure that some of the oxygen tanks stored on site had exploded but I didn't see for myself. Someone yelled that we all needed to evacuate and I was good with that. But I was severely alarmed because as I was leaving I was told and I could see that several pipes had cracked open, including what I believe were cold water supply pipes. That would mean that coolant couldn't get to the reactor core. If you can't sufficiently get the coolant to the core, it melts down. You don't have to be a nuclear scientist to figure that out.

On the way to his car, he saw reactor No. 1's outside walls starting to crumble. 'There were holes in them,' he said. 'In the first few minutes, no one was thinking about a tsunami. We were thinking about survival.'

Yet another worker was on his way into work when the earthquake happened, well before the tsunami burst in:

I was in a building nearby when the earthquake shook. After the second shock wave hit, I heard a loud explosion that was almost deafening. I looked out the window and I could see white smoke coming from reactor 1. I thought to myself, 'This is the end.'

Shortly after, a manager ordered everyone to evacuate: 'There's been an explosion of some gas tanks in reactor 1, probably the oxygen tanks. In addition to this there has been some structural damage, pipes have burst, meltdown is possible. Please take shelter immediately.'

The point is that all this happened long *before* the tsunami struck. It was the earthquake that started the nuclear disaster at Fukushima, not, as the official story would have us believe, a tsunami that no one could predict. Without adequate coolant, Fukushima's broken nuclear reactors were a serious danger long before the seawater burst in to complete the destruction.

The first indications that the nuclear beast had escaped from its cave was when an early-warning radiation alarm suddenly went off a kilometre downwind from reactor No. 1 *before* the tsunami hit the coastline and the plant. TEPCO later suggested that the alarm might have been faulty and simply malfunctioned. Shortly afterwards, the Pacific Tsunami Warning Center issued a code red tsunami warning and many of the power station workers fled to the top floor of a building near the site and waited to be rescued.

Forty-two minutes after the earthquake, huge waves 10–30 metres high (estimates vary) burst on to the Japanese coast, sweeping away vehicles, causing buildings to collapse, and severing roads and railways. The Japanese government declared a state of emergency for the nuclear power plant near Sendai, 290 kilometres from Tokyo. Sixty to seventy thousand people living nearby were ordered to evacuate to shelters.

Those tsunami waves also travelled across the Pacific, reaching Alaska, Hawaii and even the far-away Norwegian fjords. The wave was still around two metres high when it reached the shore in Chile, 16,090 kilometres away. In Japan, the towering wall of seawater created havoc as it burst into the power plant. The results were catastrophic. Seawater, electricity and red-hot nuclear reactors do not mix well.

According to *The Chunichi Shimbun* newspaper extremely high levels of radiation were already being measured within the reactor No. 1 building by mid-afternoon on 11 March. The levels were so high that just one day's exposure would have been fatal. The coolant water levels of the reactor were also becoming alarmingly low. By 9:51 pm that evening, TEPCO's Chief Operations Officer ordered the reactor building to be declared a radioactive no-entry zone. By around 11:00 pm on the evening of the earthquake, radiation levels inside the turbine building, next door to the reactor, had reached potentially lethal hourly levels of 0.5–1.2 mSv/rads. The meltdown was already underway.

At 7:03 pm that day, a Japanese Nuclear Emergency had been declared, and at 8:50 pm the Fukushima Prefecture issued an evacuation order for people within two kilometres of the plant. At 9:23 pm, Prime Minister Naoto Kan extended this to three kilometres. Early next morning, he revised the figure up to ten kilometres (six miles). He visited the plant soon after and then doubled the evacuation zone to twenty kilometres.

Although TEPCO later insisted that the cause of the meltdown was the tsunami knocking out emergency power systems, at the 7:47 pm TEPCO press conference the same day their spokesman claimed that the emergency water circulation equipment and reactor core automatic cooling systems would work even without electricity. Events were to prove that, even as he briefed the journalists, the reactor core was already cooking off and about to go critical.

Around dawn on 12 March, it did just that. The cooling water levels were now far too low and the radiation levels began rising along with the temperature. Meltdown was taking place. The TEPCO press release issued at 4:10 am on 12 March blandly stated, 'The pressure within the containment vessel is high but stable.' There was an ominous sentence buried in the press release that many harassed journalists missed, however: 'The emergency water circulation system cooling the steam within the core has ceased to function.'

Some time between 4:00 pm and 6:00 pm on 12 March, Masao Yoshida, the plant manager, decided it was time to pump seawater into the reactor core – fast. It would contaminate and wreck the runaway reactor permanently, but it would also cool it down and prevent an explosion. Tooru Hasuike, employed by TEPCO from 1977 to 2009, and formerly the Fukushima plant's general safety manager, subsequently observed: 'The emergency plans for a nuclear disaster at the Fukushima plant had no mention of using seawater to cool the core. To pump seawater into the core is to destroy the reactor. The only reason you'd do that is if no other water or coolant was available.'

There was, however, a sudden inexplicable delay. Seawater was not pumped in until two hours later, after a hydrogen explosion at about 8:00 pm on the evening of 12 March. Best guess is that in true Japanese management style, Yoshida was checking with his bosses that his decision was 'correct'. By then it was too late. The reactor had gone critical and was in meltdown.

However, there is another possible explanation for the delay. The crisis in Japan exposed the unspoken secret relationship between nuclear energy and nuclear war. Nuclear energy has never been just a civilian economic activity. It is also an appendage of the nuclear weapons industry, because nuclear reactors are the main producers of the fission by-products needed as key components for nuclear bombs. The powerful governmental interests behind nuclear energy and nuclear weapons overlap and always have.

In Japan at the height of the disaster, 'The nuclear industry and government agencies [were] scrambling to prevent the discovery of atomic-bomb research facilities hidden inside Japan's civilian nuclear power plants,' according to another investigative journalist, Yoichi Shimatsu. Writing in *Global Research* a month later, he claimed to have uncovered evidence suggesting that the Japanese governing elite had been secretly manufacturing and stockpiling components for a future nuclear weapon at Fukushima. In an article published on 12 April, and provocatively titled 'Secret Weapons Program Inside Fukushima Nuclear Plant?', Shimatsu claimed that consulting the government in Tokyo was the real reason for the crucial delays at Fukushima that evening when the reactor crashed and began to melt down.

Back at the Daiichi plant on 12 March, the drama continued. Reactor No. 2 lost all its coolant and began to cook off. Officials quickly began to pump seawater into the reactor, as they had been doing with two other reactors at the same plant. Workers scrambled to cool down fuel rods at reactors Nos. 1 and 3. A fresh explosion at No. 3 then caused a building wall to collapse, injuring six. The 600 residents remaining within thirty kilometres of the plant, despite an earlier evacuation order, were now ordered to stay indoors.

On 15 March, there was a third explosion at the Daiichi plant, which affected reactor No. 2's suppression pool; to cool radioactive material, water had to be injected into 'pressure vessels'. The following day, a white cloud of smoke above the plant was investigated by the nuclear safety agency; it was later established as vapour from a spent-fuel storage pool. Meanwhile, the Emperor Akihito addressed the population in a rare television broadcast: 'We need to understand and help each other.' Two days later the INES level was raised from 4 to 5, with the warning that there was a probability of radioactive material being released, serious damage to the reactor core and the likelihood of some deaths from radiation sickness. Then, on 12 April, the Japanese nuclear agency raised the crisis to the highest level, 7, denoting 'a major accident', as serious as the 1986 disaster at Chernobyl, and now threatening both health and environment.

Away from the plant, panic set in among Japanese citizens.

Thousands had been killed and hundreds of thousands displaced and made homeless. At least 6 million homes, or 10 per cent of Japan's households, were without electricity, and a million were without water. Some 9,500 people, half the town's population, were reported to be unaccounted for in Minamisanriku on Japan's Pacific coast. Although the Japanese Nuclear and Industrial Safety Agency announced that radiation on the plant's perimeter was more than eight times the normal level, there was no government information issued about the increase in the region's radiation levels, meaning that people could not be clear about their own safety. A general panic set in. Even Red Cross workers fled. Local doctors were in the dark. 'We had no information for ten days,' remembered Koichi Hasegawa, a consultant at Fukushima Medical University. Food and other supplies from Tokyo for survivors in the stricken region went undelivered as no one was prepared to drive into what they saw as a radioactively 'hot' environment.

Over the coming months, the details of the drama emerged slowly. On 6 June 2011, Japan's Nuclear Emergency Response Headquarters confirmed that reactors Nos. 1, 2 and 3 at the Fukushima Daiichi nuclear power plant had gone critical and experienced a total meltdown. By the end of that month, the Japanese government announced that more households within fifty to sixty kilometres of the power plant should be evacuated while government teams were checking radiation levels. In July, Kansai Electric's Ohi nuclear plant shut down due to difficulties in the emergency cooling system – so now only eighteen out of Japan's fifty-four nuclear plants were supplying the country with electricity. At the end of October, in a curious attempt to reassure the public, government official Yasuhiro Sonoda proposed to demonstrate the safety of contaminated water by drinking some, live on television.

Looking back at the Fukushima disaster, it stands as a clear and stark warning of the inherent dangers of nuclear power.

One of those is that governments automatically lie and try to downplay nuclear disasters, just as they did at Windscale and Chernobyl. It is the default position of officialdom. At Fukushima, the reason for official reluctance to admit that the earthquake did direct structural damage to reactor No. 1 only became obvious after the disaster.

In his book *TEPCO: The Dark Empire* (2007), Katsunobu Onda explains: 'If TEPCO and the government of Japan admit an earthquake can do direct damage to the reactor, this raises suspicions about the safety of every other similar reactor. They are using a number of antiquated reactors that have the same systematic problems, the same wear and tear on the piping.'

Onda notes, 'I've spent decades researching TEPCO and its nuclear power plants and what I've found, and what government reports confirm, is that the nuclear reactors are only as strong as their weakest links, and those links are the pipes.'

During his research, Onda spoke with several engineers who worked at the TEPCO plants. Several questions were raised. Did the piping match the blueprints? Was the inspection rigorous? Did the design allow for easy access? Were the repairs carried out by TEPCO engineers or subcontractors? Onda believes that it is not very difficult to explain what happened at Fukushima reactor No. 1 and probably at many others as well. 'It was like a maze of pipes inside,' he explains. 'The pipes, which regulate the heat of the reactor and carry coolant, are the veins and arteries of a nuclear power plant; the core is the heart. If the pipes burst, vital components don't reach the heart and thus you have a heart attack, in nuclear terms.'

Onda was not alone in his warnings. Kei Sugaoka, a GE on-site inspector who worked at reactor No. 1, had notified Japan's nuclear watchdog, the Nuclear Industrial Safety Agency (NISA) in June of 2000 that TEPCO had deliberately falsified safety records. The Japanese government did not address these issues for over two years, and the identity of the whistleblower was leaked.

Sugaoka identified cracks in reactors Nos. 1 to 5 in the Fukushima Daiichi Power Plants. The cracks in the pipes resulted from general long-term usage. In September 2002, TEPCO admitted to covering up data concerning cracks in critical circulation pipes in addition to previous falsifications.

Just nine days before the meltdown of 11 March 2011, NISA issued a final warning to TEPCO over its failure to check the plant's most crucial equipment, notably the coolant re-circulation pumps. TEPCO was given three months to inspect and repair accordingly. Mr Sugaoka said that he wasn't surprised that a meltdown took place after the earthquake:

> The plant had problems galore and the approach taken with them was piecemeal. Most of the critical work: construction work, inspection work, and welding were entrusted to subcontracted employees with little technical background or knowledge of nuclear radiation. I can't remember there ever being a disaster drill.

The accident had wider and longer-term consequences. Very low levels of radioactive chemicals that leaked from Fukushima have been detected along the North American coast in Canada and California. Trace amounts of radioactive isotopes caesium-134 and caesium-137 have been found in seawater collected in 2015 and 2016.

The Japanese nuclear plants that were not immediately damaged were gradually shut down. For nearly two years between mid-2013 and mid-2015, Japan suspended nuclear power generation for the first time in more than forty years. Fukushima itself has been entombed.

'While Chernobyl was an enormous unprecedented disaster, it only occurred at one reactor and rapidly melted down. Once cooled, it was able to be covered with a concrete sarcophagus that was constructed with 100,000 workers. There are a staggering 4,400 tons of nuclear fuel rods at Fukushima, which greatly dwarfs the total size of radiation sources at Chernobyl,' according to Global Research in an article titled 'Extremely High Radiation Levels in Japan: University Researchers Challenge Official Data' (11 April 2011).

Japan recovered slowly. Starting at the end of 2015, electric utilities received the necessary approvals to recommission a few reactors. The disaster broke the country's longstanding commitment to nuclear power and prompted a four-year moratorium on the country's atomic energy production. Public opposition to post-Fukushima nuclear power and delays in the governmental approval process have brought the whole future of Japanese nuclear power generation into question. In some coastal regions only 58 per cent of people headed for higher ground immediately after the earthquake in 2011, assuming the incoming tsunami would be as small as those they had previously experienced. Japan's Meteorological Agency now has a new, upgraded tsunami warning system.

The direct financial damage from the disaster is estimated to be about $200 billion dollars (about 17 trillion yen), according to the Japanese government. The total economic cost could reach up to $235 billion, according to the World Bank.

That makes Fukushima's earthquake and tsunami the most expensive natural disaster in history – so far.

CHAPTER 24

FALSE ALARMS

'Whether it is an accident, terrorism or irresponsible push of a nuke button, we all know the result is devastating. How to prevent accidental nuclear explosions on earth? This is the most crucial political, moral, social, technical and spiritual question of our time.'
– AMIT RAY, *PEACE ON THE EARTH: A NUCLEAR WEAPON FREE WORLD* (2017)

On 13 January 2018, 'Alert!' signs popped up on mobile phones all over Hawaii and traffic signs warned of incoming missiles. Horrified Hawaiians read:

BALLISTIC MISSILE THREAT INBOUND TO HAWAII.
SEEK IMMEDIATE SHELTER.
THIS IS NOT A DRILL.

Sirens wailed. Thousands panicked, thinking that Kim Jong-un had finally gone berserk and North Korea had launched nuclear rockets against America's island state. Fortunately, within minutes, new signs appeared reassuring citizens that this actually was just a drill.

Scary as this erroneous emergency notification was, it was far from the first such false alarm America had faced. During the Cold War, governments dealt regularly with hundreds of anomalies and errors that could have led to a nuclear launch. The Hawaii event didn't even reach the military's chain of command or decision-makers in government.

Within three minutes of the 2018 alert, the Hawaii National Guard had contacted the US Pacific Command and the Honolulu Police Department and advised them that it had been a false alarm. The incident had been sparked when a supervisor at the Hawaii Emergency Management Agency (HI-EMA) was running an unscheduled drill during a shift change. It appears that one employee had (not for the first time) 'confused real life events and drills'. Convinced that a real emergency was taking place, he followed his training procedures. Despite his repeated apologies for the misunderstanding, he has since been the recipient of numerous death threats.

Official embarrassment was not confined to Hawaii. Only three days later, Japanese broadcaster NHK sent a similar false alert to 300,000 followers of its 'Japanese News and Disaster Prevention' service concerning a North Korean missile fired at Japan. Public reaction was muted, as within five minutes of the notification NHK admitted that it was a mistake.

A similar incident happened in Canada two years later. On 12 January 2020, the Ontario emergency operations centre mistakenly issued an immediate nuclear alert to all television and radio stations. The message was swiftly confirmed as a false alarm and a mistake during a routine internal test; an operations officer had accidentally forgotten to log out of the live alert network.

What these alerts demonstrate is the ever-present danger of a false alarm caused by human error. A nuclear close call is nothing more than a perceived imminent nuclear threat that invites instant retaliation by the targeted nuclear-armed country. When this is just a panicky warning to civilians to take cover, the damage is limited, as in Hawaii, but when it confuses an opponent's own warning system into believing that nuclear weapons are on their way, the potential for an accidental war is much more serious. The prospect of nuclear war is bad enough, but nuclear war because 'someone blundered' would be even more tragic. If such mistakes and false alarms were isolated affairs, we could be relatively relaxed. Unfortunately, history shows that they are distressingly frequent.

Though exact details on many nuclear close calls are hard to come by, analysis of known cases highlights three common factors. First, most are caused by human error. Second, technical breakdowns (in one case a malfunctioning microchip costing less than a dollar) are frequent. And last, it is vital to have a human supervisor in the loop to decide what is real and what is false. Artificial intelligence and complex algorithms are no substitute for a cool head and experience in a nuclear crisis.

Both the main protagonists in the Cold War, the USA and the USSR, recognised the need for an emergency communications system during a crisis, real or imagined, and it was no surprise when, after the nail-biting days of the Cuban stand-off in autumn 1962, a hotline was set up between Washington and Moscow. The politicians, particularly in Washington, also realised that they need to corral their aggressive military leadership and began to insist on tighter technical control, and locking procedures to minimise the possibility of a nuclear war starting without their say-so or as a result of a false alarm.

A direct hotline can be vital, as situations can be misunderstood very easily during a time of increased international tension. A classic example occurred in November 1956 when British and French forces attacked Egypt, prompted by the latter's nationalisation of the Suez Canal earlier that year. The Soviets invited the USA to cooperate and prevent such an offensive through joint military action, also issuing warnings to the British and French governments to back away. Meanwhile, reports emerged that the crisis was worsening:

- Unidentified aircraft were reported flying over Turkey and the Turkish Air Force was on alert.
- 100 Soviet MIG-15s were identified flying over Syria.
- A British Canberra bomber had been shot down over Syria.
- The Soviet Black Sea fleet was deploying into the Mediterranean through the Dardanelles.

The truth turned out to be somewhat different:

- The unidentified aircraft over Turkey was a large flight of swans.
- A small ceremonial Soviet squadron was escorting the President of Syria on his way home from a very public visit to Moscow.
- An RAF Canberra bomber had been forced to make an emergency landing because of mechanical problems.
- The Soviet fleet was setting out on a long-scheduled and widely advertised routine exercise.

If it hadn't been so serious it would have been funny. Swans as the cause of World War III? Those Suez alerts of 1956 were a prime example of misread intelligence. Perhaps more ominous is the fact that a war could start because of a false alarm caused by a mistake in the warning systems. That alert would automatically put the 'threatened' party's nuclear retaliation weapons on readiness. The supposed aggressor would note this in turn, and then put their own nuclear forces on alert.

That initial false alarm would have started a chain reaction of nuclear nervousness. All it then needs to start a nuclear exchange, and a war, is a hair trigger or some rogue military man jumping the gun. Even an accidental civilian nuclear explosion could have the same result, especially during a period of increased international tension.

Many false alarms and accidents occurred during the Cold War. Many remain unknown to the public. Some of the false alarms are positively bizarre. In one case, during the days when early long-range radars looked out to the horizon, the ascending moon in the sky was misinterpreted as a missile attack. In another, a fire on a broken gas pipeline was reported as the deliberate jamming, by laser, of a satellite's infrared sensor.

The 'Rising Moon' false alarm of 5 October 1960 has become particularly famous as an example of technical equipment fooling human operators. Radar equipment in Thule, Greenland, mistakenly went on red alert as they recorded a huge wave of Soviet ICBMs coming in over Norway. North American Aerospace Defense Command (NORAD) declared an immediate alert. The somewhat calmer press corps picked up on this and quite reasonably enquired why on earth Moscow would want to nuke New York City at a time when Soviet Premier Nikita Khrushchev was in the city leading the USSR's United Nations delegation? Puzzled diplomats agreed. NORAD checked again and admitted somewhat shamefacedly that the 'incoming Soviet strike' was, in fact, the moon rising over Norway.

Nerves are, understandably, more twitchy in times of real tension. During the Cuban Missile Crisis of 1962, small mistakes, even by individuals, were easily magnified into something bigger and much more serious. On the night of 27 October 1962, one of the tensest days of the crisis, U-2 pilot Captain Charles Maultsby was ordered to fly a reconnaissance mission to the North Pole; he could only navigate by star shots and sextant. Unable to take accurate readings, however – because of the atmospheric conditions caused by the aurora borealis, or 'Northern Lights' – he accidentally strayed into Soviet airspace. MIG interceptors were scrambled with orders to shoot him down. The alarmed American fled east towards Alaska, pursued by the Russian planes. A pair of US F-102 fighters were then launched to escort him and prevent the MIGs from entering US airspace. A simple mistake by a confused pilot nearly became an international stand-off and was wrongly interpreted by the Kremlin as a deliberate escalation, causing increased international tension.

If that was downright dangerous, what had happened two nights earlier was just plain farcical. An alert guard at the Duluth Air Defense Sector (DADS) in Minnesota spotted a dark figure trying to clamber over the security fence. He shot at it, then activated his 'sabotage alarm'. This automatically set off sabotage alarms at other bases in the area. At Volk Field, Wisconsin, nuclear-armed F-106 jets prepared for take-off. The pilots believed that World War III had started, because at the DEFCON 3 stage of defence readiness there are no practice drills or sorties. Everything is for real. Duluth then called up and said it was all a mistake. The jets were stopped by a car screeching on to the runway as they tried to take off. It turned out that the alert sentry had shot a hungry bear.

Wandering bears apart, the tense atmosphere of the Cuban Missile Crisis gave rise to one very serious false alarm indeed. Just before 9:00 am on 28 October 1962, in Moorestown, New Jersey, radar operators informed the national command post that a nuclear attack was underway. It wasn't. Against standard operating procedures, a test tape simulating a missile launch from Cuba was being run during a real DEFCON 3 situation, and just at the same time a real satellite was picked up routinely coming over the horizon. Unfortunately, the Moorestown radar post had not received information about the routine satellite transit because the station responsible had been given other missions for the duration of the DEFCON crisis.

Moorestown operators became confused and reported by telephone to the NORAD headquarters that an ICBM impact was expected thirty kilometres west of Tampa, Florida, at 9:02 am. A disbelieving NORAD broadcast an alert but it was too late to react – no action could be taken in time. At 9.04 am, the telephone from Tampa rang, with a puzzled officer querying, 'What's all this about a nuclear detonation? Ain't nuthin' goin' on down here. Is this some kinda drill?'

Embarrassed Moorestown USAF operators reported a false alarm. Later investigation uncovered that, during the incident, overlapping radars that should have confirmed or disagreed on events were offline and oblivious to any alert. Once again, human error had overridden technology.

As we have seen (see pages 216–20), on 27 October 1962 off Cuba, the world came within a whisker of a nuclear exchange, not caused by the popular misconception of ICBM missiles and nuclear bombers but by tired, frightened men at sea. This, along with 'Exercise Able Archer' in 1983, has to have been one of the most dangerous military nuclear alarms of the Cold War era.

Mechanical failure plays its part all too often in false nuclear alarms. Electricity itself is sometimes the problem. On 9 November 1965, the Command Center of the US Office of Emergency Planning were dealing with a serious power cut in the northeastern United States, when several nuclear bomb detectors appeared to announce a nuclear attack. The slightly more mundane truth was that the bomb detectors, designed to differentiate between regular power outages and those outages caused by nuclear blasts, malfunctioned or sent false messages because of circuit errors.

Sometimes technical devices are fooled by natural phenomena. In May 1967, solar researchers noticed a big group of sunspots with strong magnetic fields. On 23 May, the sun fired off a flare so powerful that it was visible to the naked eye, and began emitting radio waves at a level that had never been seen before. Intense flares hit Earth and began disrupting radio transmissions, satellite communications and sensitive long-range radars.

In North America, this was wrongly interpreted as deliberate Soviet jamming of all three of the Air Force's Ballistic Missile Early Warning System radar sites in Alaska, Greenland and the United Kingdom – an obvious prelude to war. The US Air Force began moving nuclear assets to increased readiness. Fortunately, military space observers intervened in time and the nuclear strike preparations were quietly stood down. The 1967 'sun flare story' became well known to the public and was even the source of some mocking cartoons.

Not so the 'unknown' nuclear crisis of 1969. It was not until 2010, when Michael Gerson and colleagues at the Center for Naval Analyses' Strategic Studies Department published an astonishing paper on a much-overlooked nuclear crisis of the Cold War, the Sino-Soviet clash of 1969, that people realised just how close the Far East had come to a nuclear war. And it was more a near-miss than a false alarm.

In early 1969, Beijing and Moscow were engaged in localised armed clashes over the old 1860 Tsarist border on the Ussuri River. The situation had begun to get out of hand and eventually came to a full nuclear alert as the Chinese government secretly fled Beijing and deployed its nuclear forces for war. The Russians had inflamed tensions by sending signals intended to warn the Chinese off, even asking a US State Department official what Washington's reaction would be if 'the Soviet Union attacked and destroyed China's nuclear installations'. Kissinger correctly identified this as 'a brutal warning in an intensified war of nerves'. Yet when CIA Director Richard Helms leaked that the Soviets had been approaching Communist governments in Eastern Europe about an attack on China's nuclear programme, Mao and the Chinese government were seriously spooked and issued orders for war.

In their turn, the Soviets were becoming anxious about a

Chinese conventional autumn attack on Siberia. According to a defector, 'The Politburo was terrified that the Chinese might make a large-scale intrusion into Soviet territory... A nightmare vision of invasion by millions of Chinese made the Soviet leaders almost frantic. Despite our overwhelming superiority in nuclear weaponry, it would not be easy for the USSR to cope with an assault of this magnitude.'

The situation was getting out of control. False alarms and misconceptions on both sides threatened a serious nuclear crisis as Chinese decision-makers convinced themselves that a Soviet sneak attack was imminent. Increasingly bellicose Soviet diplomatic statements prompted Mao to transfer elite military units and air force surface-to-air missile battalions from the south to the north; new tank divisions were formed; new air-raid shelters were built throughout the country and Beijing ministries even began to evacuate critical archives. The Central Committee of the Communist Party of China (CCCPC) advised all Communist Party, military and civilian leaders to leave Beijing before 20 October and brought nuclear weapons to full readiness. Mao himself fled to Wuhan in central China.

Both sides then seem to have come out of some diplomatic trance and to have recognised that things were escalating out of control.

On 20 October 1969, after eight months of violence and threatening political rhetoric, China and the Soviet Union finally sat down at the negotiating table. It took several years to hammer out a new agreement concerning the border between Siberia and the People's Republic. Beijing claimed on more than one occasion that its bargaining position was disadvantaged because 'above the negotiating table hangs the Soviet atomic bomb'. But for the Chinese leadership, and the Kremlin, the threat of a nuclear strike and a nuclear war had been very real.

This combination of international tension and operator error creating false alarms seems almost inevitable. On 24 October

1973, a technician's mistake nearly started a nuclear exchange between the USSR and the United States during the Egypt-Syria-Israel Yom Kippur War. It certainly started a panic at the Kincheloe Air Force Base in Michigan.

US intelligence had good evidence that the USSR was planning to move into the war zone and help the Egyptians. President Nixon was busy struggling to escape the toils of Watergate, so Secretary of State Kissinger ordered DEFCON 3 – purely as a precaution. The purpose of the alert was not to prepare for war, but to warn the Soviets not to intervene in the Sinai and to leave it to the Israelis to deal with the Egyptians and the Syrians. The next day, 25 October 1973, electricians at a tense and wary Kincheloe Air Force Base were repairing one of the warning klaxons and accidentally set off the whole base alarm system. Hearing the klaxons, B-52 bomber crews at DEFCON 3 rushed out to their bombed-up, nuclear-armed aircraft and started the engines. The duty officer in the control tower quickly realised it was a false alarm and recalled the aircrew before they could take off. A calm, alert human brain in the loop stopped a potential disaster.

Technology of a more advanced kind than a simple klaxon nearly caused one of the biggest false alarms of the Cold War. At 8:50 am on 9 November 1979, NORAD spotted a pattern indicating a large number of inbound Soviet missiles. US senator Charles Percy was visiting NORAD headquarters that day and claimed that there was absolute panic at the news. Emergency preparations for nuclear retaliation were swiftly made. A number of air force planes were launched, including the US National Emergency Airborne Command Post, although the US President, who may have been in the White House lavatory at the time, was not aboard. For some unexplained reason no one tried to call the Kremlin on the hotline.

NORAD then checked with its separate Precision Acquisition Vehicle Entry Phased Array Warning System (PAVE PAWS), a group of coastal-based 'phased array' early-warning radars; a phased array is more effective than a conventional rotating radar transmitter because it can track multiple targets using individual electronic pulses. PAVE PAWS reported no missiles, nor were the sensors on the surveillance satellites in space showing any sign of incoming missiles. In only six minutes the panic was over and the threat assessment conference ended. It turned out that yet again the villain was an exercise tape running on the live computer system. Questions were asked in Congress and in future all US early-warning exercise tapes were run from a system not linked to live military hardware.

Phased array radars are huge and very expensive. Not so are the basic components in a conventional rotating radar. At 2:25 am on 3 June 1980, NORAD warning displays at their command centres started showing large numbers of missiles detected. This was odd but, as a precaution, preparations for retaliation were instigated: nuclear bomber crews started their engines, Pacific Command's Airborne Command Post was launched and Minutemen missiles were readied for launch. Experienced radar operators, however, were convinced that this was a false alarm and so it transpired.

An investigation was started but three days later the same thing happened and preparations for retaliation were restarted. It turned out that the cause was a single faulty computer chip costing 46 cents that was failing erratically and, because the basic design of the computer system was faulty, that glitch was causing identical errors across several command posts. US State Department advisor Marshall Shulman later stated that 'false alerts of this kind are not a rare occurrence. There is a complacency about handling them that disturbs me.' In the months following the incident there were three more false alarms at NORAD, two of them caused by faulty computer chips.

Technical failures and faulty radars were not confined to the USA.

The USSR had suffered several worrying false alarms, but none as dangerous as the incoming American ICBMs scare of 26 September 1983. The incident took place in the weeks after the downing of Korean Air Lines Flight 007 over Soviet airspace and tensions were still running high when a satellite early-warning system near Moscow reported the launch of a small number of Minuteman ICBMs. By sheer chance, Lieutenant-Colonel Stanislav Petrov, who had been one of the designers of the Soviet Air Defence System, was duty officer that night when the system alerted him to a missile attack from America. The alarms went off as the warning system's computer registered that a missile had been launched from a base in the United States.

On the panel in front of Petrov was a flashing red button marked 'Start'. He hesitated. But Petrov was employed to analyse the incoming data, and he was forced to make a split-second decision. Should he press the button? When the alarms went off, on the night of the crisis, Petrov had little time to think. He recalled, 'For fifteen seconds, we were in a state of shock. We needed to understand, what's next?' He said later that he knew that something was wrong with the systems. 'I knew it could not be an incoming attack because there were only five missiles. Pointless.'

Petrov's position was critical. Technically, he reported to colleagues at warning-system headquarters, who in turn consulted with general staff in Moscow, staff who would then discuss launching a retaliatory attack with Soviet leader Yuri Andropov. Normally, a single report of a small-scale rocket launch did not immediately go up the chain to the general staff and the Krokus electronic command system. Despite the electronic evidence, Petrov decided, based on his own experience, that the satellite alert of the ICBMs was a false alarm. He sat on the report.

In doing so, he probably averted a nuclear holocaust.

'I had a funny feeling in my gut,' Petrov told American journalist David Hoffman of the *Washington Post* in February 1999. 'I didn't want to make a mistake. I made a decision, and that was it.' Warned many times of the enormity of a nuclear attack, he was well aware that just one would instantly and singularly overpower any and all Soviet defences. 'When people start a war, they don't start it with only five missiles. You can do little damage with just five missiles.'

Initially, once the false alarm had been confirmed, Petrov received much praise – until a KGB security investigation took place. He was asked why he had not written everything down that night. His response was: 'Because I had a phone in one hand and the intercom in the other, and I don't have a third hand.' The only official recognition he eventually received was a reprimand for failing to keep an accurate log during the incident itself. In the eyes of Soviet bureaucrats, saving the planet from Armageddon was no excuse for not doing the paperwork. It turned out that the false alarm was caused by a satellite that picked up the sun's reflection off the tops of clouds and mistook it for a missile launch.

Petrov was dismissed from the service a year later and lived on a paltry pension in a run-down Moscow suburb. The official account is still considered secret by the Russian authorities. When Kevin Costner heard the story, he celebrated Petrov in a 2014 film as *The Man Who Saved the World*. Petrov enjoyed a few years of international celebrity before he died at the age of seventy-seven in 2017.

Even a presidential joke prompted an alert in those tense last days of the Cold War. On 11 August 1984, preparing for his regular Saturday afternoon radio broadcast, President Ronald Reagan quipped in a microphone test that he had 'signed legislation that will outlaw Russia forever' and that 'we begin bombing in five minutes'. His little joke was recorded and the recording promptly leaked, to be broadcast, much to his embarrassment, on news networks around the world. Months later, *The Times* reported that two days after President Reagan's joke, a Soviet military official had ordered an alert of troops in the Far East. The Kremlin ordered it stopped immediately and American intelligence officials later briefed that the alert was 'a non-event'.

Even after the Cold War, suspicions died hard. A particularly incompetent Soviet false alarm happened on 25 January 1995, when Russian early-warning radars detected an unexpected missile launch near Spitzbergen. The estimated flight time to Moscow was five minutes. The Russian President Boris Yeltsin, the Defence Minister and the Chief of Staff were informed. Yeltsin became the first world leader to activate a nuclear briefcase and his launch codes. The Russian early-warning system and the control and command centre switched to full combat mode. Then the radars determined that the missile's impact would be outside the Russian borders. It turned out that the missile was a harmless Norwegian Black Brant XII research rocket being used to study the Northern Lights. Worse for the Kremlin, on 16 January, Norway had notified thirty-five countries, including Russia, that the peaceful rocket launch was planned. Information had apparently reached the Russian Defence Ministry but they had failed to pass it on to their early-warning radar system. It was worse than human error: it was incompetence.

Taking stock of these mishaps, it is hard to assess the true level of risk. One thing is clear. False nuclear alarms have been a fact of life for decades. One day, one is bound to go its full course, whether from miscalculation, equipment malfunction, or human error – with disastrous results.

CHAPTER 25

THE SUM OF ALL FEARS: TERRORISTS WITH A NUKE

'Nuclear terrorism is one of the most serious threats of our time. Even one such attack could inflict mass casualties and create immense suffering and unwanted change in the world forever. This prospect should compel all of us to act to prevent such a catastrophe.'
– BAN KI-MOON, UN SECRETARY-GENERAL (2007–16)

Westerners are obsessed with battles. Napoleon and the Duke of Wellington have much to answer for. While the crushing defeat of the French emperor near Waterloo in 1815 was the final blow to the old European order, Napoleon – and the great Prussian military theorist and general Carl von Clausewitz – fostered a very dangerous misunderstanding: the myth that it was great, decisive battle that won wars.

That simply isn't the case, especially in the modern world. The decisive head-to-head clash on the battlefield is by no means the only way for mankind to fight. For the weak and the ruthless, sometimes the stiletto in the back has been the only way to attack powerful enemies. Terrorism is as old as time, and attacks on the rich and powerful have a long and bloody history as old as humanity itself. The modern-day notion that terrorism is some kind of a new phenomenon is totally incorrect.

There are two main kinds of terrorism: 'terror tactics' and 'terrorism' itself. Terror tactics are deliberately horrific acts of mass murder that have been used throughout recorded history to send an enemy the message: 'Don't resist us, or this will happen to you.' The Romans' total destruction of Carthage and Genghis Khan's mounds of skulls are a brutal reminder of atrocity as an act of war.

They were by no means the first, nor the last, acts of terror designed to intimidate. But terror in war is not terrorism. The purpose of terror in peacetime is to intimidate and frighten – even your own population. The French, the Russians, the Chinese and the Cambodian people have been on the receiving end of a domestic terror meted out by their own government, determined to control its own citizens by cowing them. After the Bolshevik seizure of power in 1917, Felix Dzerzhinsky – the architect of its secret police, the Cheka – spelled it out very clearly: 'We Chekists represent organised terror; nothing else...'

Dzerzhinsky's targets were those opposed to the Red Revolution at home or abroad, but terrorism nowadays tends to mean a specific act of violence against a selected target, usually to obtain some political advantage. The phrase 'armed propaganda' is entirely accurate. Terror attacks are intended to communicate a clear threat, whether in pursuit of a grievance or to warn enemies off. The aim is always to frighten, but at the same time to send a message to three key constituencies: the enemy, the domestic population and, lastly, to overseas observers.

Terrorism has been around in various forms for centuries. Jewish fanatics known as *sicarii* plunged curved knives into the backs of off-duty legionaries in the dark alleys of Jerusalem during the Roman occupation of first-century Palestine, and political murder has always been with us, as Guy Fawkes's 1605 Gunpowder Plot demonstrates.

Closer to our own time, the murderous Thugs of India, travelling gangs who strangled fellow travellers in the name of their bloodthirsty god, Kali, and the brutal way the Mafia traditionally held sway in Sicily are both examples of terrorism in action, dominating whole populations by the use of fear as a weapon.

The Industrial Revolution gave the terrorist three invaluable new tools: accurate, quick-loading firearms, reliable explosives and a political theory. This political blueprint really comes out of Tsarist Russia. In the last half of the nineteenth century, a number of revolutionary groups began to target politicians and officials in an attempt to bring down the Tsarist state through a deliberate campaign to assassinate and terrorise its functionaries – an attack on the power of the state by killing its officials and by terrifying its population.

The question a traumatised public is entitled to ask is, why? What do today's terrorists want? Most people nowadays point to religion, and specifically fanatical Muslim jihadis. But the question then arises, what political advantage are these jihadi 'warriors for God' trying to achieve? Here, the waters of terrorist calculation and logic suddenly become very murky. What do modern terrorists flying aircraft into the World Trade Center really want? To convert the rest of the world to Islam? Kill all the non-believers, the *kafir*? Bring about the end of world? It's not always entirely clear. Sometimes it looks like sheer atavistic hatred of the West and everything it stands for.

There is no doubt that Islam bears a strong sense of grievance against the West. The roots of hostility lie deep. In the 500 years after the Prophet Muhammad's death in AD 632, Islam developed an advanced scientific and literary culture. Mathematics, astronomy, medicine, geography, music, poetry, botany and metallurgy all flourished at a time when most Europeans lived in ignorance. Western attempts to set up Christian crusader kingdoms in Palestine were stopped short by Salah ad-Din (Saladin), who recaptured Jerusalem in 1187. For Islam this was a golden age, but eventually the caliphate began to contract. In 1492, Columbus reached the Americas, but it was also the year that the Spanish finally expelled the last 'Moors' (Muslims) from Spain. Only the Ottomans kept the flame of the faith alive for the next two hundred years, until they were defeated at the gates of Vienna in 1683.

By the early nineteenth century, the sultans were in general retreat as successful European commerce and expansionism invaded large swathes of the Islamic world, succeeding where the Crusaders' force of arms had failed nine centuries before. The result was that by the middle of the twentieth century, discontented, broken, resentful Islam had a serious grievance against the new Western invaders. Islamic militancy was reborn as nationalism.

Yet when the nationalist Arab leaders such as Nasser of Egypt, Assad of Syria and the reinstalled Persian Shah of Iran finally freed themselves from the so-called 'yoke of colonialism', they discovered that their radical Muslim clerics were far from grateful. Organisations like the Muslim Brotherhood and the Mosque became a permanent source of trouble. The new regimes cracked down hard on their dissident priests. In Shi'ite Iran, opponents of the Shah either fled to avoid prison or were exiled. Chief among these Iranian exiles was an obscure ayatollah, or 'Leader of the Shi'a', called Khomeini. A clear fault line was emerging between church and state in the Islamic world.

To these religious leaders of Islam, paradise was not some secular utopia on Earth, with freedom, liberty and equal opportunities for all, votes for women and well-stocked supermarkets. For radical Islam, the true path should be based on an Islamic ideal and the revealed word of God. Put simply, during the last years of the twentieth century radical Muslim clerics proclaimed a gospel of hope to fire up a new generation of postcolonial Muslim youth and imbue them with a single burning slogan: 'The Koran is our constitution.' Better still, Sunni and Shi'a could now unite against a single God-given enemy of all Islam: Israel.

In 1979, the Iranian Revolution overthrew the Shah. Allah – and his messenger on Earth, Khomeini – now ruled Iran and the Muslim Shi'a world. Within the year, Iran was officially declared an Islamic republic and all secular opposition crushed. Khomeini called on Muslims everywhere to fight against the Shah's backer – the 'Great Satan' of the West, America – and its servant, Israel. Among the teeming millions of the Arab and Islamic world, where 50 per cent of the population is under the age of thirty and dirt poor, this Iranian call to arms struck home. In Lebanon, Palestine, Gaza and the West Bank, disaffected young Arabs, both Sunni and Shi'a, rallied to jihad, or what they saw as 'holy war'. The Palestine Liberation Organisation began a series of spectacular terrorist attacks, hijacking planes to publicise Palestinian grievances, while Hezbollah, a terror group backed by Iran, invoked the call of martyrdom, carrying out a series of suicide bombings against Western targets. Gradually, terrorism and Islam became fused in the minds of many outside observers. It seemed that Islamic terrorism was everywhere.

These terrorists now had a cause, they had a seemingly inexhaustible supply of enthusiastic recruits and they had more than enough money from oil to buy basic weapons. Moreover, the long-running war against the 'godless' Communist Soviets in Afghanistan had given them sophisticated new weapons, combat experience and organisational skills. By the mid-1980s, the Great Satan's CIA was actually arming the Afghan Islamic Mujahideen in a proxy war to fight the Soviet Union. After the Soviets gave up and pulled out of Afghanistan in 1989, one man stepped out of the rubble determined to continue the conflict as a global 'holy war', a jihad against the USA and its allies. Osama bin Laden's organisation, al-Qa'ida, ('the Base') would become a sworn enemy of the West. What had begun as a logistic network to support Muslims fighting against the Soviet Union continued, but now recruiting, arming and training a new generation of jihadi terrorists to carry the war by extending its targets to include the West and what they saw as corrupted Islamic regimes.

Bin Laden's group absorbed a number of other militant Islamist organisations, established camps for Muslim militants from throughout the world, training tens of thousands of recruits in paramilitary skills, and engaged in numerous terrorist attacks, including the destruction of the US embassies in Nairobi, Kenya, Dar es Salaam, Tanzania and a suicide bomb attack against the USS *Cole* in Aden. On 11 September 2001, al-Qa'ida staged the spectacular 9/11 aerial attacks on the USA itself, causing nearly 3,000 deaths when airliners ploughed into the twin towers of the World Trade Center in New York and two other US locations, one of them the Pentagon itself.

Osama bin Laden's motives remain obscure to this day, as do the roots of his fanaticism and even his underlying aim. The wellsprings of his grievance and hatred have never really been clear. What exactly was he trying to achieve? What is obvious is that he was bent on causing mayhem by using bloody terrorism as al-Qa'ida's signature calling card. Bin Laden swiftly built up a wide-reaching organisation claiming that an unholy Christian–Jewish alliance was conspiring to destroy Islam and must be confronted and destroyed.

The terrorist organisation concentrated on four main tasks: organising a web of financial backers; absorbing sympathetic co-respondent organisations; building a network of international support workers; and recruiting warriors capable of finding global soft targets and new weapons. The true motives of al-

Qa'ida and its successor, the Islamic State of Iraq and Syria (ISIS), may remain broad and ill-defined but they plainly included driving US armed forces out of the Middle East and East Africa. At bin Laden's request, sympathetic Muslim clerics obligingly issued religious *fatwahs* (rulings handed down by an Islamic scholar or leader) stating that such attacks were both sanctioned by God and necessary to build Allah's Dar al-Salam – the Islamic 'House of Peace'.

For terror groups intent on wreaking maximum destruction on a hated enemy, acquiring nuclear weapons of some sort was an early priority. The bigger the bang, the greater damage and more publicity, runs the terrorist thinking. Nuclear weapons have been the holy grail of the jihadi godfathers from the beginning. With one device, they could wreak havoc, fire and slaughter on millions. The most dangerous terrorist of all is the one with nuclear arms. Nuclear terrorism is nowadays defined as the sabotage of a nuclear facility and/or the detonation of a radiological device, colloquially termed a 'dirty bomb'. In legal terms, nuclear terrorism is an offence committed if a person unlawfully and intentionally 'uses in any way radioactive material . . . with the intent to cause death or serious bodily injury; or with the intent to cause substantial damage to property or to the environment; or with the intent to compel a natural or legal person, an international organisation or a state to do or refrain from doing an act,' according to the 2005 United Nations International Convention for the Suppression of Acts of Nuclear Terrorism.

America recognised the danger very early. In December 1945, as a new Cold War loomed, worried congressmen quizzed the 'father of the atomic bomb', J Robert Oppenheimer, about the possibility of smuggling atomic weapons into the United States. Oppenheimer acknowledged the danger, but said that tracking illicit movement would be difficult, because every incoming packing case would need to be opened and inspected.

This unhelpful reply sparked further work on the question of smuggled atomic devices during the 1950s, including small 'suitcase bombs'.

Attempts to steal nuclear materials attracted greedy crooks, not just would-be terrorists. In 1966, twenty natural uranium fuel rods were stolen for their scrap value from Bradwell nuclear power station in the UK by Harold Arthur Sneath, a worker at the plant. The local police quickly recovered the rods when they stopped a clapped-out van, driven erratically by Sneath's gormless accomplice. The pair hadn't realised the radioactive danger and were bound over and fined. History doesn't relate whether they subsequently glowed green in the dark.

Other, more serious, incidents highlighted the vulnerability of nuclear facilities. In November 1972, hijackers took over a US domestic passenger flight and threatened to crash the plane into a US nuclear weapons plant in Oak Ridge, Tennessee. The plane was as close as 2,440 metres (8,000 feet) above the site before the hijackers' demands for money were met and the plane flew on to Cuba. There was even a failed attempt in March 1973 by a domestic terrorist group to take over the still incomplete Atucha I Nuclear Power Plant in Argentina.

As the Cold War hardened and nuclear weapons developed, the topic of nuclear terrorism and nuclear vulnerability became a matter of worried debate among experts. *The Economist* warned in 1975 that:

> You can make a bomb with a few pounds of plutonium. By the mid-1980s power stations may easily be turning out 200,000 lb [90,720 kg] of the stuff each year. And each year, unless present methods are drastically changed, many thousands of pounds of it will be transferred from one plant to another as it proceeds through the fuel cycle. The dangers of robbery in transit are evident... Vigorous co-operation between governments and the International

Atomic Energy Agency could, even at this late stage, make the looming perils loom a good deal smaller.'

In 1981 the *New York Times* explicitly warned of the dangers of nuclear weapons in the wrong hands, pointing out:

> [The Nuclear Emergency Search Team's] origins go back to the aftershocks of the Munich Olympic massacre in mid-1972. Until that time, no one in the United States Government had thought seriously about the menace of organised, international terrorism, much less nuclear terrorism . . . But it has since been revealed that the physical safeguarding of bomb-grade material against theft was almost scandalously neglected.

This was strong stuff, but shows that by the 1970s the dangers were recognised and understood. Matters were not helped by the fact that instructions for making a nuclear device were freely available in the public domain and later on the internet. In 1983, when America's NBC aired *Special Bulletin* – a made-for-TV movie about a speculative terrorist attack on the USA in which a homemade nuclear device is detonated by terrorists, destroying much of Charleston, South Carolina – public awareness resulted in pressure on the US government for action.

In 1986, the International Task Force on the Prevention of Nuclear Terrorism advised that nuclear-armed states needed to do more to guard against dangers of nuclear terrorism. Experts warned that the likelihood of nuclear terrorism was growing, adding 'The consequences for urban and industrial societies could be catastrophic.'

Other groups such as the World Institute for Nuclear Security (which was formed after an attempt in South Africa to steal weapons-grade uranium) and the Global Initiative to Combat

Nuclear Terrorism – founded in 2006, this is a group of nations, originally thirteen, now eighty-nine, working together on preventing, detecting and reacting to nuclear terrorist incidents – warned governments that the threat of nuclear terrorism was escalating.

The possibility of terrorist organisations acquiring and using nuclear weapons is now openly discussed as plausible and possible. But despite such an alarming prospect, and notwithstanding the thieving and trafficking of fissile Category III Special Nuclear Material, no terrorist group – it appears – has yet obtained any Category I SNM, the kind and volume of weapons-grade plutonium needed to make a nuclear weapon. This does not mean the threat has gone away. Many US experts consider the eventual terrorist detonation of a dirty bomb containing radiological materials to be inevitable. 'I'm surprised it hasn't happened yet,' Nancy Holgate, the US senior advisor on nuclear terrorism, warned a Washington conference on nuclear security and the dangers of nuclear terrorism in 2013, 'because the mechanics of such a device are simple and widely known.'

Her warning was reinforced by a US Energy Department report on the nuclear terrorist threat issued that same year, which stated, 'We know that it would not require a team of nuclear physicists or even a particularly sophisticated criminal network to turn raw material into a deadly weapon ... In many cases, a determined lone wolf or a disgruntled insider is all it might take.'

We know three key facts about the threat today. First, terrorism is ongoing, widespread and will continue to be so. Just a random five-year snapshot of the more notorious terrorist attacks shows how prevalent terrorist attacks are:

- 29 March 2010 – Russia. Moscow Metro bombings; 40 dead, 102 injured.
- 1 May 2010 – USA. Attempted detonation of a car

bomb in Times Square, New York. Faisal Shahzad, the perpetrator, was affiliated with Tehrik-i-Taliban Pakistan (the Taliban in Pakistan).
- 28 May 2010 – Pakistan. Attacks on mosques in Lahore by Tehrik-i-Taliban, killing nearly 100 people and injuring many others.
- 18 July 2011 – China. Hotan attack. A group of 18 young Uyghur men perpetrated a series of coordinated bomb and knife attacks and occupied a police station.
- 5 January 2012 – Iraq. Bombings, in Baghdad and Nasiriyah, by ISIS, Islamic State of Iraq; 73 dead, 149 injured.
- 3 May 2012 – Russia. Makhachkala attack; 14 dead, including 2 suicide bombers; 130 wounded.
- 11 September 2012 – Libya. Attack on the US Consulate in Benghazi; 4 dead, 11 injured.
- 15 April 2013 – USA. Boston bombings. Two brothers planted two bombs near the finish line of the Boston Marathon; 3 killed, 183 injured.
- 11 May 2013 – Turkey. Reyhanlı bombings, killed 52 people and wounded 140 others.
- 22 May 2013 – UK. Two men with cleavers butcher British soldier Lee Rigby on the street in Woolwich, London.
- 21 September 2013 – Kenya. Westgate shopping mall attack; 67 killed, 175 wounded.
- 22 September 2013 – Pakistan. Peshawar church attack; 80–83 killed, 250 wounded.
- 20 May 2014 – Nigeria. Jos bombings; at least 118 killed and over 56 injured.
- August 2014 – Iraq. Islamic State fighters massacred some 700 people, mostly men, of the Shu'aytat tribe.
- 23 September 2014 – Australia. Numan Haider, an Afghan Australian, stabbed two counter-terrorism officers in Melbourne, Australia.

- 22 October 2014 – Canada. Shootings at Parliament Hill, Ottawa. Lone attacker shot a soldier at a war memorial and attacked Parliament.
- 28 November 2014 – Nigeria. Kano bombing. Around 120 people were killed and another 260 injured.
- 16 December 2014 – Pakistan. Peshawar school attack. Over 140 people dead, including at least 132 children.
- 18 December 2014 – Syria. Mass grave of 230 tribesmen murdered by Islamic State found in eastern Syria.
- 18 December 2014 – Nigeria. Gumsuri kidnappings. Boko Haram insurgents killed 32 people and kidnapped at least 185 women and children.
- December 2014 – Iraq. Islamic State militants execute 150 women, some of whom were pregnant at the time, who refused to marry their fighters.
- 25 December 2014 – Somalia. Al-Shabaab attack in Mogadishu leaves 9 dead.
- 28–29 December 2014 – Cameroon. Boko Haram attacks a village, leaving 30 dead.
- 7 January 2015 – France. Paris. Coordinated attacks killing 130 and wounding 430, including those murdered at offices of French satirical magazine *Charlie Hebdo*.

This is just a selection of the worst cases. There were at least three times as many more terrorist incidents recorded during this five-year period but not listed above.

From this grisly global list of terror, misery and death, a couple of thought-provoking facts emerge. Firstly, most of the victims of Muslim terror are fellow Muslims; and secondly, the fear of terrorism may be exaggerated. Over the past decade, terrorists have killed an average of 21,000 people worldwide each year. The number of Western casualties is well below deaths from other causes in modern developed societies. An

American's chance of dying from terrorism is virtually zero. He or she is much more likely to die from a shooting (15,292 people were fatally shot in the United States in 2019, excluding suicides) or a road traffic accident (36,560 deaths in 2018) than from terrorism.

It is also clear that terrorism tends to be very geographically focused. In 2017, a mighty 95 per cent of terrorist deaths occurred in the Middle East, Africa or south Asia, areas with internal domestic conflicts. An American is more likely to die from old age, falling over, opioid overdoses (40,000) or suicide (38,000) – both figures from 2018 – than being slaughtered by a terrorist. The truth is that media coverage of terrorism is often disproportionate to its frequency and the terrorist death toll.

This, of course, fails to dispel the unique fears of a terrorist nuclear attack. Terrorist attacks appear to be made at random. Then there is the question of scale. Even a small 'dirty bomb' could contaminate huge areas for a generation or more. That alone is genuine cause for alarm. And any nuclear detonation would kill or injure thousands – perhaps millions – at a single stroke.

Another factor contributing to our fear of this kind of event is the sheer, unremitting determination of the terrorists to get their hands on a bomb by hook or by crook. The most likely sources are obviously where the nuclear material is being stored. A number of vulnerable locations offer cause for concern. Chief among these is nuclear-armed, and fundamentally Islamic, Pakistan. The country has a small stockpile of nuclear bombs and much of its Muslim population is openly sympathetic to the jihadis' cause. It is the obvious first place to look. Building your own bomb from scratch and in secret requires access to fissile ingredients and is not easy, even for developed nations with well-qualified scientists and well-equipped laboratories. Better to acquire one 'ready to go'.

According to a 2011 report published by Harvard's Belfer

Center for Science and International Affairs, terrorists looking for a nuclear capability have only a limited number of options:

- Being secretly given a device by a sympathetic group inside an existing nuclear state.
- Getting a nuclear weapon that has been stolen or bought on the black market.
- The acquisition of fissile material from a sympathetic nation-state.
- The use of a crude explosive device built by rogue nuclear scientists recruited or blackmailed through money, ideology, compromise or ego.
- The use of an explosive device constructed by terrorists and their accomplices using their own fissile material.

Of these options by far the most dangerous, and the most likely, are the first two. That places Pakistan firmly in the witness box.

In 2009, a West Point Military Academy paper alleged that Pakistan's nuclear sites had been attacked by al-Qa'ida and the Taliban at least three times. Pakistan's military rejected these allegations as 'absolute nonsense'. Later information, however, indicated that there had been three nuclear incidents – suicide attacks aimed at causing maximum damage, rather than seizing weapons. The US Army now allegedly has an undercover special forces team with the sole task of neutralising and recovering any lost Pakistani nuclear weapons. This seems prudent, as the West Point paper also found that Pakistan's stockpile 'faces a greater threat from Islamic terror groups seeking nuclear weapons than any other nuclear stockpile on Earth'. In 2016, the US Defense Intelligence Agency blandly reported that Pakistan 'continues to take steps to improve its nuclear security, and is aware of the threat presented by extremists to its program'.

Not everyone shares this confidence. Rolf Mowatt-Larssen,

a former CIA investigator, warned that there is 'a greater possibility of a nuclear meltdown in Pakistan than anywhere else in the world. The region has more violent extremists than any other, the country is unstable and its arsenal of nuclear weapons is expanding.'

David Albright, a nuclear weapons expert and author of *Peddling Peril: How the Secret Nuclear Trade Arms America's Enemies* (2010), was not reassured either. He believes that Pakistan 'has had many leaks from its program of classified information and sensitive nuclear equipment, and so you have to worry that it could be acquired in Pakistan.'

These fears are compounded by today's bloodthirsty jihadis, still hell-bent on acquiring a nuclear capability. In his first speech to the UN Security Council, US President Barack Obama warned that 'Just one nuclear weapon exploded in a city, be it New York or Moscow, Tokyo or Beijing, London or Paris, could kill hundreds of thousands of people.' Nuclear terrorism would 'badly destabilise our security, our economies and our very way of life . . . and is the single most important national security threat that we face.'

That al-Qa'ida and its supporters still wish to acquire nuclear weapons is not in doubt. Leaked diplomatic documents found in bin Laden's hideouts indicate that the jihadi movement is working hard to produce radiological weapons, simply by obtaining nuclear material and recruiting rogue scientists to build devices containing radioactive contaminants that can be blown into the atmosphere by conventional explosives – a kind of poor man's cheap-and-cheerful dirty bomb. A terrorist attack using such a makeshift explosive is 'a nightmare waiting to happen,' says Frank Barnaby, a nuclear consultant who used to work at the UK's atomic weapons plant in Aldermaston in Berkshire. 'I'm amazed that it hasn't happened already.'

This seems to be what al-Qa'ida and ISIS are now working towards. They have been attempting to purchase nuclear weapons and have sought nuclear expertise on numerous occasions for over two decades without much luck. How much easier just to buy some stolen nuclear material or radioactive waste and let it off where it can do the most harm? The only confirmed case of attempted nuclear terrorism so far involved a dirty bomb. On 23 November 1995, Chechen separatists buried a crude bomb containing 32 kilograms of a mixture of caesium-137 radioactive isotopes and dynamite in Moscow's Izmailovsky Park. Reporters were tipped off about its location, however, and it was defused.

Before he was killed, Osama bin Laden even stated that the acquisition of nuclear weapons or other weapons of mass destruction was a 'religious duty'. And there is no indication that the wider ISIS and jihadi movement has abandoned its attempts to acquire a bomb or some fissile material. For Rolf Mowatt-Larssen, there are three potential headlines that especially worry him and presumably most responsible politicians: terrorists with Pakistani 'loose nukes', terrorists with nuclear bombs supplied by North Korea and a nuclear attack by al-Qa'ida or ISIS. The good news is that he thinks the likelihood of terrorists acquiring a nuclear bomb in the near future are slim. In fact, by 2019 the number of terrorist attacks worldwide had halved since 2006. In 2006, there were 14,371 terrorist attacks, while by 2018 only 8,093 attacks were recorded. The statistics show the number of deaths due to terrorism worldwide had dropped as well. In 2017, terrorism was responsible only for 0.05 per cent of global deaths. This can be attributed mainly to the collapse of ISIS, as it lost its territory in Iraq and Syria; when it came under growing territorial pressure, the group transitioned back to insurgent operations, concentrating on preserving its original core capability. When the attacks did strike, they were 'lone wolf' spectaculars, such as the individual behind the bombing of a concert by Ariana Grande in Manchester in 2017, vehicle attacks on crowds in

Nice, Germany and Barcelona by supporters of the group and the coordinated attacks in Paris in 2015 that killed 130 people and wounded 430 others.

But this decreased tempo and intensity of terrorism has to be weighed against the terrible consequences of any well-planned and precisely targeted nuclear atrocity. This was all-too-convincingly depicted in the 2002 film *The Sum of All Fears*, in which a small nuclear device in a packed sports stadium kills thousands of spectators, including the US president.

In an article for the *Bulletin of the Atomic Scientists* in 2010, Rolf Mowatt-Larssen noted that 'a probability-based approach to managing risk' makes less sense than an approach 'focused on mitigating threats in descending order of their possible consequences'. Cutting through the academic jargon, what this nuclear expert is saying is that the risk of terrorism is low, but the effects of a nuclear attack don't bear thinking about. The message is clear: just a single nuclear terrorist attack would have devastating consequences.

And there are those who are, today, looking to do just that.

CHAPTER 26

FUTURE UNCERTAIN? SUMMARY AND CONCLUSIONS

'The probability of complete human extinction by nuclear weapons is at 1% within the century, the probability of 1 billion dead at 10% and the probability of 1 million dead at 30%.'
– SURVEY FROM THE 'GLOBAL CATASTROPHIC RISK CONFERENCE' (2008) AT THE FUTURE OF HUMANITY INSTITUTE, OXFORD UNIVERSITY

The original purpose of this book was to examine the consequences of the splitting of the atom. Like many of life's journeys, its travels and destination have not been as straightforward as was anticipated. We have encountered some eyebrow-raising moments along the way, from Japan's surprising enthusiasm for things nuclear to naive scientists' unthinking curiosity. But the facts and the deductions we can draw are clear to all. In our attempts to prepare for the future, we can hopefully learn some lessons and draw some conclusions from events in the past.

Nuclear power in all its forms is always seriously dangerous. Like an idiot juggling hand grenades, we treat nuclear devices far too casually, out of greed or for profit, or for their enormous military destructive applications. Secondly, our knowledge about the energy locked within the atom cannot be disinvented, it can only be controlled. Last, nuclear devices carry a long-term alchemist's curse that is ignored or glossed over at our peril. Radioactivity is the real danger and it lasts a very long time.

Overlying the whole subject and the debate are the politics of the issue, with emphasis on the most troubling topic of all – how to think about nuclear weapons and their use. This was neatly summed up by John Woodcock, ex-special advisor to the British MOD and a former Member of Parliament. Addressing the responsibilities of any prime minister or national leader, he made a statement in December 2019 with regard to Britain's nuclear weapons and deterrence that makes the point succinctly and more clearly than any reams of official government policy statements or academics' self-serving, jargon-laden papers:

> If you effectively totally disable your nuclear deterrent from day one of being prime minister, because you have been open that you would never use it under any circumstances – by definition, it is no longer a deterrent. And that places not only this country but also all of those other countries that currently fall within the NATO umbrella at greater risk of nuclear blackmail.

His logic, although aimed at a British domestic audience, can be applied to any country possessing nuclear weapons or contemplating acquiring them, not just NATO. Nuclear weapons bring power, a wary respect, deter any would-be aggressor bent on blackmail and carry a wider logic of their own. Possession does, however, demand a willingness to use them, if only in the hope that that particular 'Devil's Alternative' will never be tested in anger. Capability has to be accompanied by the threat of intention.

Another factor to be considered is the question of the basic usefulness of weapons of mass destruction. During World War II, both sides built stocks of lethal chemical weapons – from 1944 onwards, German chemists even developed nerve agents such as VX and sarin, although they were never used. The reason lies in a self-denying ordinance, fears of retaliation and a mutual Armageddon. Both sides chose to be deterred by these deadly weapons. Hitler's experience of being gassed in the trenches during World War I undoubtedly contributed to the decision. Therein lies a lesson for nuclear weapons. We cannot disinvent them, but we can certainly control their use. It is no wonder that the words 'nuclear war' horrify people. They conjure up a host of terrors: the destruction of our lives and civilisation; the insidious, invisible radioactivity; nuclear winter and the end of our world.

Nuclear power also brings risks. Civilian accidents are all too common, with exploding power stations that spread death and destruction, poisoning the food chain and water supply, causing cancer, malformed babies and radioactive waste that remains hazardous for 30,000 years. Civilian nuclear activities cause all manner of hazards throughout the life cycle of a nuclear power source, from mining and processing to meltdowns and waste. Even mining uranium is a dangerous business and its history is appalling. During World War II, the US government began digging for uranium throughout the American Southwest to find the uranium needed to build their first atom bombs. Officials didn't warn the miners or the people living in the surrounding communities that the work was hazardous to their health. After all, bureaucrats reasoned, were they not working on a top-secret weapon?

Meanwhile, people living downstream drank water full of radioactive isotopes that seeped out of the mines. Mysterious cancer clusters began to emerge. For twenty years, no one informed the mining communities what was happening.

Politicians refused to provide the money needed to clean up the hazards. Men coughed up their lungs and died digging uranium in the 1940s and 1950s. Nowadays, the US government has comprehensive clean-up plans, and mining today is much safer, but a uranium miner still receives more radiation than any X-ray technician.

Once the uranium is mined, we cannot separate nuclear power from nuclear war. After uranium ore is milled into concentrated 'yellowcake' powder, it has to be enriched by being spun in centrifuges to transform it into nuclear fuel. The longer it spins, the more it achieves the alchemist's dream, transmuting power-plant quality uranium into different elements, and producing the unstable fissile explosive needed for a bomb. This enrichment process is the crucial connection between atomic energy and weapons, between peace and war. It is very easy, once you have a nuclear power station, to turn the uranium into the enriched fissile material needed for a bomb. Iran, for instance, has insisted that it is only enriching uranium for reactors, but the fact that it built a secret enrichment plant merely confirms it is trying to produce a bomb.

Once it's enriched, nuclear fuel goes to the reactor. That's where we find the very real and all-too-prevalent dangers of accidents and meltdowns. When the reactor at Pennsylvania's Three Mile Island partially melted down in 1978, no one was killed, and there was only a small release of radiation. But the accident came within a whisker of a nuclear explosion and hundreds of thousands evacuated their homes in a mass panic.

Worst of all the nuclear reactor accidents was the botched test at the Chernobyl plant in the Soviet Ukraine in 1986. The nuclear reactor exploded through the roof of the building around it. People ninety-five kilometres (sixty miles) away felt the ground shake. Of the three high-profile accidents since nuclear plants started running in 1951, Chernobyl was the worst.

Then there's the Fukushima meltdown, which caused no

direct fatalities but was an unparalleled disaster in terms of cost and damage. A 2017 report from the United Nations concluded that health effects to the general public from radiation were almost zero. But the earthquake, subsequent tsunami and panicked evacuation of 110,000 people led to 1,600 deaths. The truth is, nuclear disasters are devastating and terrifying. They capture the attention of the world – and with good reason.

And finally, there remains the problem of nuclear waste material, enriched or otherwise. We currently produce around 400,000 tons of nuclear waste globally. Nuclear waste is not the glowing green gloop beloved of the cartoons. It consists of solid pellets or long metal rods holding spent uranium. The rods go into a pool of water, and then, when radioactivity has cooled off, into metal and concrete containers. They then have to be buried somewhere, deep underground in caves made of 1.5-metre-thick (4-foot-thick) reinforced concrete and steel to keep the radiation in for centuries. Let us hope that they stay there.

What does this depressing tale of accidents, secrecy, lies, government cover-ups and, worst of all, miscalculation, reveal? It warns us that, like it or not, nuclear is here to stay; it is forever dangerous in all its forms and accidents are inevitable. There is, however, a glimmer of hope.

Because of their sheer awfulness, nuclear weapons may – just may – be becoming semi-obsolete. As the nature of warfare changes, so the reliance on brute force and mass destruction becomes less attractive as a means of prosecuting 'politics by other means', to misquote von Clausewitz. New, more subtle methods of waging war are emerging. Hybrid warfare, cyber attacks, biological and other deniable options to head-to-head confrontation are becoming more attractive as alternatives to kinetic weapons and firepower. And they carry less risk of nuclear retaliation. To the modern strategist, a deniable cyber attack that disables an adversary's power grid, computers and electronic systems, or some new biological virus that wrecks a whole society and its economy, are much more attractive options than invading or bombing and risking nuclear retaliation. The new weapons' sheer deniability becomes attractive.

How can a nuclear strike be launched when there is no clear and confirmed target? Strategic nuclear weapons become redundant in the face of such an offensive, except for what Henry Kissinger called 'spasm' or revenge warfare. This does not make nuclear completely redundant. The controlled use of tactical weapons at sea, in the air or to halt an otherwise unstoppable incursion are still useful tools of war, but vaporising an aircraft carrier is an act of war of a very different order of magnitude to obliterating Beijing and all its inhabitants.

Writing in 2020, Edward Lucas, a respected and experienced writer on security and defence, identified the changing tools and face of warfare:

> China is also gaining the edge in artificial intelligence, robotics, life sciences and space technology. Unlike the US, it does not burden itself with burdensome defence spending and foreign wars. It buys cheap 'carrier-killer' missiles, each of which costs less than one thousandth of the giant warships that America believes vital for its global prestige.

If he is correct, much of the world's strategic nuclear arsenal risks becoming redundant, an expensive but unusable luxury item gathering dust on a shelf.

None of this takes away from the dangers of nuclear accidents. The evidence is plain to see. As long as nuclear weapons and nuclear power stations exist, the danger from the caged atom will remain. So, at some point in the future, based on what we have uncovered and experienced so far, we can confidently forecast another disastrous nuclear accident, military or civil, brought about by a reactor meltdown, an accidental weapon discharge or a natural disaster – or even a rogue bunch of terrorists or lone, mad general bent on causing maximum mayhem.

We are perched on the eve of destruction, perhaps more by accident than by design. At the very least, we face serious danger ahead or some unimaginable disaster, one way or another, from our perilous atomic tools and weapons.

We have had plenty of warning.

APPENDIX:

NUCLEAR-ARMED COUNTRIES WITH WEAPONS-GRADE HIGHLY ENRICHED URANIUM, AND COUNTRIES WITH NUCLEAR POWER

Countries with nuclear arsenals:

    United States of America   India
    Russian Federation   Pakistan
    United Kingdom   North Korea
    France   Israel
    China

Countries formerly with nuclear weapons:

    South Africa   Kazakhstan
    Belarus   Ukraine

Countries with nuclear power:

    Argentina   Bangladesh*
    Armenia   Belarus*

Belgium	Romania
Brazil	Russia
Bulgaria	Slovakia
Canada	Slovenia
China	South Africa
Czech Republic	South Korea
Finland	Spain
France	Sweden
Germany	Switzerland
Hungary	Taiwan
India	Turkey*
Iran	Ukraine
Japan	United Arab Emirates
Mexico	United Kingdom
Netherlands	United States of America
Pakistan	

Note: at the time of writing, countries here marked with an asterisk had nuclear reactors under construction but not yet operational.

NOTES AND SOURCES

This book was written primarily for the interested general reader and not as an academic textbook.

If more detail or further reading is required, then I have found the following books and other references to be especially useful or interesting, particularly those giving direct access to contemporary primary sources, such as the various CIA documents, Russian accounts or Sir William Penney's highly selective official report on Windscale, for example. Without Google and Wikipedia's invaluable lists of primary references, I suspect that I would still be wandering around searching the shelves of university libraries...

I also agree strongly with my colleague Max Hastings, that lengthy bibliographies are sometimes more evidence of some kind of academic 'virility parade' than any real help to those looking for sensible guidance for further reading. For this reason I have confined myself only to those books and sources that I felt had most to offer, rather than every single possible reference. Life, and the reader's attention span (let alone the reader's budget), is just too short...

Introduction

Samuel Upton Newtan, *Nuclear War 1 and Other Major Nuclear Disasters of the 20th Century*. AuthorHouse, 2007.

Paul Voosen, 'Hiroshima and Nagasaki Cast Long Shadows over Radiation Science', *E&E News*, 11 April 2011. https://www.eenews.net/stories/1059947655.

Samuel Glasstone and Philip J Dolan (eds.), *The Effects of Nuclear Weapons* (rev.). United States Department of Defense, 1962.

'Nuclear Bomb Effects': The Atomic Archive. Undated.

Douglas Holdstock and Frank Barnaby (eds.), *Hiroshima and Nagasaki: Retrospect and Prospect*. Frank Cass, 1995.

Toshiko Marks, Ian Nish and R John Pritchard, 'Japan and the Second World War', *International Studies Paper Series 197*. Suntory and Toyota International Centres for Economics and Related Disciplines, LSE, 1989.

Paul Ham, *Hiroshima, Nagasaki: The Real Story of the Atomic Bombs and Their Aftermath*. Picador, 2015.

Chapter 1 Atoms

Carl Wentrup, 'Marie Curie, Radioactivity, the Atom, the Neutron, and the Positron', *Australian Journal of Chemistry* 64 (7), August 2011.

Jack Challoner, *The Atom: A Visual Tour*. MIT Press, 2018. An accessible and engaging guide to the atom.

Lawrence M Krauss, *Atom: An Odyssey from the Big Bang to Life on Earth*. Abacus, 2002.

Niels Bohr, 'The Structure of the Atom.' Nobel Lecture, 11 December 1922.

'Periodic Table of the Elements', Husted, Robert; et al. Los Alamos National Laboratory. December 2003. Good for all the isotopes.

Jeremy Bernstein, 'A Memorandum That Changed the World', *American Journal of Physics* 79(5), May 2011.

Francis George Gosling, *The Manhattan Project: Making the Atomic Bomb*. United States Department of Energy, 1994.

Richard H Campbell, *The Silverplate Bombers: A History and Registry of the Enola Gay and Atomic B-29s Configured to Carry Atomic Bombs*. McFarland and Co. Inc., 2005.

Chapter 2. Ruthless Rays: Radiation – Silent but Deadly

Timothy J Jorgensen, *Strange Glow: The Story of Radiation*. Princeton University Press, 2016.

'Radiation Dose', *Factsheets & FAQs: Radiation in Everyday Life*. International Atomic Energy Agency (IAEA), 2013.

Eileen Welsome, *The Plutonium Files: America's Secret Medical Experiments in the Cold War*. The Dial Press, 1999. United States government secret experiments on unwitting Americans.

'Report of the United Nations Scientific Committee on the Effects of Atomic Radiation to the General Assembly', UNSCEAR, 2008.

Denise Grady, 'A Glow in the Dark, and a Lesson in Scientific Peril', *New York Times,* 6 October 1998.

'Acute Radiation Syndrome: A Fact Sheet for Clinicians (Physician's Fact Sheet)', CDC Radiation Emergencies Fact Sheets, 22 April 2019. https://www.cdc.gov/nceh/radiation/emergencies/arsphysicianfactsheet.htm.

Chapter 3. Oops! Some Early Accidents

Sarah Watt, 'From Marie Curie to the Demon Core: When Radiation Kills', *Discover Magazine*, 23 March 2018. https://www.discovermagazine.com/health/from-marie-curie-to-the-demon-core-when-radiation-kills.

David Irving, *The German Atomic Bomb: The History of Nuclear Research in Nazi Germany*. Da Capo, 1983.

Eileen Welsome, op. cit.

Alex Wellerstein, 'Demon Core and the Strange Death of Louis Slotin', *New Yorker*, 21 May 2016. https://www.newyorker.com/

tech/annals-of-technology/demon-core-the-strange-death-of-louis-slotin.

Matt Reimann, 'The First Fatal Nuclear Meltdown in the U.S. Happened in 4 Milliseconds', *Timeline*, 19 December 2016. https://timeline.com/arco-first-nuclear-accident-f16ec1105b9c.

Lynn Vincent and Sara Vladic, *Indianapolis: The True Story of the Worst Sea Disaster in U.S. Naval History and the Fifty-Year Fight to Exonerate an Innocent Man*. Simon and Schuster, 2018.

Chapter 4. Some Early Military Near-misses

Richard Rhodes, *The Making of the Atomic Bomb* (25th Anniversary Edition) Simon and Schuster, 2012.

David Holloway, *Stalin and the Bomb: The Soviet Union and Atomic Energy, 1939–1956*. Yale University Press, 1994.

Meyers K Jacobsen, *Convair B-36: A Comprehensive History of America's 'Big Stick'*. Schiffer Publishing Ltd, 1998.

Dirk Septer, *Lost Nuke: The Last Flight of Bomber 075*. Heritage House Publishing, 2016.

'Details of 1957 H-Bomb Accident Told by Crewmembers', Associated Press News, 30 August 1986. https://apnews.com/62b9d336e7a0fdb7f693800a4b896982.

Michael H Maggelet and James C Oskins, *Broken Arrow: The Declassified History of U.S. Nuclear Weapons Accidents*. Lulu.com, 2008. Details on the thirty-six known Broken Arrows.

Sergey A Zelentsov, 'Memoirs of Lieutenant-Colonel N. V. Danilenko' in Ядерные испытания 'Nuclear Exercises' Vol. II. Картуш, 2006.

'Military List of Accidents': CPEO Archive. Includes rebuttal from Defence.

Chapter 5. Balls-up at Bikini

'Operation Crossroads: Bikini Atoll', *Naval History and Heritage Command*. https://www.history.navy.mil/our-collections/art/

exhibits/conflicts-and-operations/operation-crossroads-bikini-atoll.html. See also: 'Operation Crossroads: Fact Sheet'. *Naval History and Heritage Command*, 2015. https://www.history.navy.mil/research/library/online-reading-room/title-list-alphabetically/o/operation-crossroads/fact-sheet.html.

Connie Goldsmith, *Bombs over Bikini: The World's First Nuclear Disaster*. Twenty-First Century Books, 2014.

Ōishi Matashichi, *The Day the Sun Rose in the West: Bikini, the Lucky Dragon, and I*. University of Hawaii Press, 2011.

'Operation Castle': Nuclear Weapons Archive, May 2006. https://nuclearweaponarchive.org/Usa/Tests/Castle.html.

Gerard H Clarfield and William M Wiecek, *Nuclear America: Military and Civilian Nuclear Power in the United States, 1940–1980*. Harper & Row, 1984.

Thomas Sumner, 'Bikini Atoll Radiation Levels Remain Alarmingly High', *Science News*, 6 June 2016. https://www.sciencenews.org/article/bikini-atoll-radiation-levels-remain-alarmingly-high.

Michael Hoffman, 'Forgotten Atrocity of the Atomic Age', *Japan Times*, 8 August 2011. https://www.japantimes.co.jp/culture/2011/08/28/books/book-reviews/forgotten-atrocity-of-the-atomic-age/#.XtJPVsDTVPY.

Chapter 6. The French Foul-up

Julian Jackson, *A Certain Idea of France: The Life of Charles de Gaulle*. Allen Lane, 2018. Described as 'the magisterial work on de Gaulle'.

Julien Peyron, 'Four Decades of French Nuclear Testing', *France24*, March 2009. https://www.france24.com/en/20090324-four-decades-french-nuclear-testing-.

'Listing Des Essais Nucléaires Français', Capcom Espace, 2005. Lists all French nuclear tests.

Campbell Page and Ian Templeton, 'French Inquiry into Rainbow Warrior Bombing', *The Guardian*, 24 September 1985.

'La Dimension Radiologique des Essais Nucléaires Français en Polynésie' (Technical report). Ministère de la Défense, Paris, March 2007. https://www.francetnp.gouv.fr/IMG/pdf/La_dimension_radiologique_des_essais_nucleaires_francais_en_Polynesie.pdf.

Philip Shenon, 'Tahiti's Antinuclear Protests Turn Violent', *New York Times*, 7 September 1995. https://www.nytimes.com/1995/09/08/world/tahiti-s-antinuclear-protests-turn-violent.html.

Chapter 7. Stop Digging! Civilian Engineering

R Dexter-Smith (ed.), *Civil Engineering in the Nuclear Industry: Proceedings of the Conference Organized by the Institution of Civil Engineers and Held in Windermere on 20–22 March 1991*. Thomas Telford, 1991.

James Lorimer, 'Some Uses of Explosives in Civil Engineering', Works Construction Paper No. 3. *Journal of the Institution of Civil Engineers* 26(5), 1946.

'Executive Summary: Plowshare Program', US Department of Energy, August 2016. https://www.osti.gov/opennet/reports/plowshar.pdf.

J G Fry, et al., 'Preliminary Design Studies in a Nuclear Excavation – Project Carryall'. *Nuclear Excavation,* 1963. 5 Reports: Presented at the 43rd Annual Meeting of the Highway Research Board January 13–17, 1964. Transportation Research Board, 1964. http://onlinepubs.trb.org/Onlinepubs/hrr/1964/50/50.pdf.

Dan O'Neill, *The Firecracker Boys: H-bombs, Inupiat Eskimos, and the Roots of the Environmental Movement*. Basic Books, 2007.

James G Speight, 'Origin and Production', *Natural Gas: A Basic Handbook* (second edition). Gulf Professional Publishing, 2019.

'On the Soviet Program for Peaceful Uses of Atomic Weapons' (PDF), Lawrence Livermore National Laboratory, 2016.

Kaushik Patoway, 'Lake Chagan, the Atomic Lake Filled with Radioactive Water', *Amusing Planet*, March 2014. https://www.

amusingplanet.com/2014/03/lake-chagan-atomic-lake-filled-with.html.

Chapter 8. The Nuclear Navy

Francis Duncan, *Rickover and the Nuclear Navy: The Discipline of Technology*. Naval Institute Press, 1989.

Thomas B Allen and Norman Polmar, *Rickover: Father of the Nuclear Navy*. Potomac Books, 2007.

'Powering the Navy National Nuclear Security Administration', NNSP official website.

Norman Polmar and K J Moore, *Cold War Submarines: The Design and Construction of U.S. and Soviet Submarines*. Potomac Books, 2004.

Brayton Harris, 'Submarine History 1945–2000: A Timeline of Development', Nova Table and Charts, *Navy Times*, 16 August 2016.

Bob Eatinger, 'Fifty-Seven Years Later: America's Worst Nuclear Submarine Disaster, *Lawfare*, April 2020. https://www.lawfareblog.com/fifty-seven-years-later-americas-worst-nuclear-submarine-disaster.

Bruce Rule, 'Assessment of Why Scorpion Was Lost by an Exceptionally Qualified Submarine Officer', IUSSCAA, 20 January 2014. http://www.iusscaa.org/articles/brucerule/assessment_of_why_scorpion_was_lost_by_an_exceptionally_qualified_submarine_officer.htm. Interesting explanation by a professional and experienced US Submariner.

Chapter 9. All at Sea: Russia's Terrible Record

Jan S Breemer, *Soviet Submarines: Design, Development, and Tactics*. Janes Information Group, 1989.

Gary E Weir and Walter J Boyne, *Rising Tide: The Untold Story of the Russian Submarines That Fought the Cold War*. Basic Books, 2003.

Carole D Bos, 'K19 Widowmaker – A Nuclear Accident', *Awesome Stories*, 2002. Retrieved February 2015. https://www.awesomestories.com/asset/view/A-NUCLEAR-ACCIDENT-K19-Widowmaker.

David Miller, *Submarine Disasters*. Lyons Press, 2006.

V P Kuzin and V I Nikolsky, *The Navy of the USSR, 1945–1991*. IMO Publications, 1996.

'Admiral Convicted in Sinking of K-159', *Moscow Times*, 19 May 2004; https://bellona.org/news/nuclear-issues/accidents-and-incidents/2004-04-northern-fleet-commander-faces-4-years-for-sinking-of-k-159.

List of All Lost Russian or Soviet Submarines. Wikipedia, 2020. https://en.wikipedia.org/wiki/List_of_lost_Russian_or_Soviet_submarines.

Chapter 10. The *Kursk* Catastrophe

Peter Truscott, *Kursk: The Gripping True Story of Russia's Worst Submarine Disasters*. Pocket Books (Simon and Schuster), 2003.

Ramsey Flynn, *Cry from the Deep: The Sinking of the Kursk, the Submarine Disaster that Riveted the World and Put the New Russia to the Ultimate Test*. HarperCollins, 2005. Strong on the Russian evasions.

'Nightmare at Sea', *St Petersburg Times*, 10 December 2004. *https://web.archive.org/web/20100823214350/http://www.sptimes.ru/index.php?action_id=100&story_id=2294.*

Miriam Pensack, 'Sinking of Russian Nuclear Submarine Known to West Much Earlier Than Stated, NSA Document Indicates', *The Intercept*, 29 May 2019.

CIA Intelligence Report, 'Russia's Kursk Disaster: Reactions and Implications', CIA Office of Russian and European Analysis, Office of Transnational Issues, 7 December 2000. Available through The Black Vault: https://documents.theblackvault.com/documents/cia/cia-kursk.pdf.

'Official four-page report on the sinking of the Kursk nuclear submarine', Russia's State Prosecutor Ustinov, *Rossiyskaya Gazeta* newspaper, 27 August 2002.

Chapter 11. Manna from Heaven? Lost Bombs

Benjamin Maack, 'A Nuclear Needle in a Haystack: The Cold War's Missing Atom Bombs', *Der Spiegel*, 14 November 2008. https://www.spiegel.de/international/world/a-nuclear-needle-in-a-haystack-the-cold-war-s-missing-atom-bombs-a-590513.html.

NORAD Agreement. North American Aerospace Defense Command, 2013. https://www.norad.mil/Newsroom/Fact-Sheets/Article-View/Article/578772/norad-agreement/. For the complete agreement see: https://2009-2017.state.gov/documents/organization/69727.pdf.

Leo Heaps, *Operation Morning Light: Terror in our Skies, The True Story of Cosmos 954*. Paddington Press, 1978.

Parker F Jones, 'Goldsboro Revisited', 22 October 1969. Declassified document in *The Guardian*, September 2013. https://www.theguardian.com/world/interactive/2013/sep/20/goldsboro-revisited-declassified-document.

Michael H Maggelet and James C Oskins, op. cit. Chapter 23.

John Megara, 'Dropping Nuclear Bombs on Spain: The Palomares Accident of 1966 and the U.S. Airborne Alert'. Florida State University, Department of History, Master of Arts thesis, 2006.

Chapter 12. Fireworks Can Be Dangerous

'The Soviet Nuclear Weapons Program'. CIA, Washington, DC, 2014.

Boris Chertok, *Ракеты и люди. Фили — Подлипки — Тюратам* (*Rockets and People: Fili – Podlipky – Tyuratam*), Mashinostroyeniye Publishing House, 1996.

Eric Schlosser, *Command and Control: Nuclear Weapons, the*

Damascus Accident, and the Illusion of Safety. Penguin Books, 2013.

August Imholtz Jr, 'The Bomarc Missile Plutonium Spill Crisis: Exercises in Propaganda and Containment in 1960 and Beyond', *The Readex Blog*, 26 April 2011. https://www.readex.com/blog/bomarc-missile-plutonium-spill-crisis-exercises-propaganda-and-containment-1960-and-beyond.

Wilson Brissett, 'Fifty-three Years Ago, Minuteman 1 Disaster Averted', *Airforce Magazine*, 21 November 2017. https://www.airforcemag.com/fifty-three-years-ago-minuteman-i-disaster-averted/.

Charles Digges, 'Fallout from Russia's Mysterious Blast Now Suggests a Reactor Blew Up, Experts Say', *Bellona*, 26 August 2019. https://bellona.org/news/nuclear-issues/2019-08-fallout-from-russias-mysterious-blast-now-suggests-a-reactor-blew-up-experts-say.

Chapter 13. Radiation Is really Bad for Your Health

J Francisco Aguirre, 'A Short Summary of Accidents Involving Radiation', PowerPoint presentation. https://international.anl.gov/training/materials/BA/Aguirre/Accidents_involving_radiation.pdf.

William R Hendee and Michael G Herman, 'Improving Patient Safety in Radiation Oncology', *Medical Physics* 38(1) and *Practical Radiation Oncology* 1(1), January 2011.

M V Ramana, 'Nuclear Power: Economic, Safety, Health, and Environmental Issues of Near-term Technologies', *Annual Review of Environment and Resources* 34, 2009.

William H Hallenbeck, *Radiation Protection*. CRC Press, 1994.

Arifumi Hasegawa, Koichi Tanigawa et al., 'Health Effects of Radiation and Other Health Problems in the Aftermath of Nuclear Accidents, With an Emphasis on Fukushima', *Lancet*, 386(9992), 2015.

William R Hendee and Michael G Herman, op. cit.

The Radiological Accident in Goiânia. International Atomic

Energy Agency, 1988. Available at: https://www-pub.iaea. org/mtcd/publications/pdf/pub815_web.pdf. See also: Diva E Puig, 'El accidente de Goiania. ¿Por qué hoy?', *Noticias Jurídicas, 2006.* https://web.archive.org/web/20140503013539/ http://noticias.juridicas.com/articulos/00-Generalidad es/200602-325612711062370.html.

'Exposure to Radiation from Natural Sources'. *Nuclear Safety & Security,* IAEA, 2016.

Chapter 14. You Can't Brush Nuclear Under the Carpet

M Ojovan and W Lee, *An Introduction to Nuclear Waste Immobilisation.* Elsevier Science, 2005.

'Hanford Overview', *Hanford Site,* United States Department of Energy, 2012. https://web.archive.org/web/20120511135540/http:// www.hanford.gov/page.cfm/HanfordOverview.

David Bodansky, 'The Status of Nuclear Waste Disposal', *American Physical Society* 35(1), January 2006.

Rob Evans, 'UK Nuclear Weapons Convoys "have had 180 mishaps in 16 years"', *The Guardian,* 21 September 2016.

'Arrangements for Responding to Nuclear Emergencies', Health and Safety Executive, 1994. Available at: http://www.onr.org.uk/ arrangement-for-nuclear-emergencies.pdf.

Michael Goulet, 'Ural Mountains Radioactivity and Health', Trade and Environment Database (TED) 7(1), January 1997. Available at: https://web.archive.org/web/20100714130850/http://www1. american.edu/projects/mandala/TED/ural.htm.

'Kyshtym Disaster', *Nuclear Heritage*, 2017. http://www.nuclear-heritage.net/index.php/Kyshtym_Disaster.

Andrew Cockburn, 'The Nuclear Disaster They Didn't Want to Tell You About', *Esquire,* 25 April 1978. https://classic.esquire. com/article/1978/4/25/the-nuclear-disaster-they-didnt-want-to-tell-you-about.

Chapter 15. Britain's Dirty Secret: The Windscale Fire

Michelle Ramadan, 'The Windscale Fire.' Introduction to Nuclear Energy (PH241), Stanford University, Winter 2016. http://large.stanford.edu/courses/2016/ph241/mjr2/.

'No Public Danger.' *West Cumberland News*, 12 October 1957 and associated national and regional press stories of the time.

Roger Highfield, 'Windscale Fire: "We Were Too Busy to Panic"', *Daily Telegraph*, 9 October 2007. https://www.telegraph.co.uk/news/science/science-news/3309842/Windscale-fire-We-were-too-busy-to-panic.html.

M J Crick and G S Linsley, 'An Assessment of the Radiological Impact of the Windscale Reactor Fire, October 1957', *International Journal of Radiation Biology and Related Studies in Physics, Chemistry and Medicine* 46(5), 1984.

William Penney, Basil F J Schonland et al., 'Official Report on the Accident at Windscale No. 1 Pile on 10 October 1957', *Journal of Radiological Protection* 37(3), 31 August 2017.

R Smallwood BSc (Hons) MIIEE, memories of Windscale. Personal communication with author, 2020.

Arthur H Wolff, 'Milk Contamination in the Windscale Incident', *Public Health Reports*, 74(1), 1959.

Fred Pearce, 'Secrets of the Windscale Fire Revealed', *New Scientist* 99(1377), 29 September 1983.

Chapter 16. The World Held Its Breath: Cuba 1962

Len Scott and R Gerald Hughes (eds.), *The Cuban Missile Crisis: A Critical Reappraisal.* Routledge, 2015.

Sergo Mikoyan, *The Soviet Cuban Missile Crisis: Castro, Mikoyan, Kennedy, Khrushchev and the Missiles of November*, ed. Svetlana Savranskaya, Cold War International History Project. Stanford University Press, 2012.

Christopher Andrew, *For the President's Eyes Only: Secret Intelligence and the American Presidency from Washington to Bush.* Harper Perennial, 1996.

'The Cuban Missile Crisis Timeline', Nuclear Files, a project of the Nuclear Age Peace Foundation, 1998. http://www.nuclearfiles.org/menu/key-issues/nuclear-weapons/history/cold-war/cuban-missile-crisis/timeline.htm.

Letter from Khrushchev to John F Kennedy, 24 October 1962. History and Public Policy Program Digital Archive, Library of Congress. https://digitalarchive.wilsoncenter.org/document/111552.

Jim Hershberg, 'Anatomy of a Controversy: Anatoly F. Dobrynin's Meeting with Robert F. Kennedy, Saturday, 27 October 1962', Cold War International History Project Bulletin 5, Spring 1995. https://www.wilsoncenter.org/publication/bulletin-no-5-spring-1995.

Leslie H Gelb, 'The Myth That Screwed Up 50 Years of U.S. Foreign Policy', *Foreign Policy*, 8 October 2012. https://foreignpolicy.com/2012/10/08/the-myth-that-screwed-up-50-years-of-u-s-foreign-policy/.

Chapter 17. The Real Cuban Missile Crisis – At Sea

'The Naval Quarantine of Cuba, 1962'. Report on the Naval Quarantine of Cuba, Operational Archives Branch, Post 46 Command File, Box 10, Washington, DC. *Naval History & Heritage Command.*

'The Submarines of October: US and Soviet Naval Encounters During the Cuban Missile Crisis'. National Security Archive, George Washington University, 31 October 2002. https://nsarchive2.gwu.edu/NSAEBB/NSAEBB75/.

Edward Wilson, 'Thank You Vasili Arkhipov, the Man Who Stopped Nuclear War', *The Guardian*, 27 October 2012. https://www.theguardian.com/commentisfree/2012/oct/27/vasili-arkhipov-stopped-nuclear-war#maincontent.

Jack A Green, 'US Navy Cuban Missile Crisis Monograph', Naval Historical Center Public Affairs. Reissued 25 February 2003.

Nikhil Gupta, *The Forgotten Heroes: Vasili Arkhipov*. CreateSpace Independent Publishing, 2017.

Chapter 18. Exercise Able Archer 1983

'The Able Archer 83 Sourcebook', Project directed by Nate Jones. National Security Archive, George Washington University. https://nsarchive.gwu.edu/project/able-archer-83-sourcebook.

'Deputy Minister Markus Wolf, Stasi Note on Meeting with KGB Experts on the RYAN Problem, 14 to 18 August 1984', 24 August 1984. History and Public Policy Program Digital Archive, Office of the Federal Commissioner for the Stasi Records. Translated for the Cold War International History Project (CWHIP) by Bernd Schaefer. https://digitalarchive.wilsoncenter.org/document/115721.

Nate Jones, 'Stasi Documents Provide Details on Operation RYaN, the Soviet Plan to Predict and Preempt a Western Nuclear Strike; Show Uneasiness Over Degree of "Clear-Headedness About the Entire RYaN Complex"', Unredacted, National Security Archive Blog, 29 January 2014. https://unredacted.com/2014/01/29/stasi-documents-provide-operational-details-on-operation-ryan-the-soviet-plan-to-predict-and-preempt-a-western-nuclear-strike-show-uneasiness-over-degree-of-clear-headedness-about-the-entire-ryan/.

Gregory Pedlow, 'Exercise ABLE ARCHER 83: Information from SHAPE Historical Files', March 28, 2013. National Security Archive, George Washington University. https://nsarchive2.gwu.edu/NSAEBB/NSAEBB427/docs/6.a.%20Exercise%20Able%20Archer%20SHAPE%20March%202013%20NATO.pdf.

Benjamin B Fischer, *A Cold War Conundrum: The 1983 Soviet War Scare*, Intelligence Monograph, Center for the Study of Intelligence, Central Intelligence Agency, 1997. Available at: https://www.cia.gov/library/readingroom/docs/19970901.pdf.

Review of Nate Jones, *Able Archer 83: The Secret History of the NATO Exercise That Almost Triggered Nuclear War*, 2016, by Jonathan M DiCicco, Humanities and Social Sciences Online, H-Diplo, August 2017. https://networks.h-net.org/node/28443/reviews/190618/dicicco-jones-able-archer-83-secret-history-nato-exercise-almost.

Chapter 19. Some Civilian Nuclear Disasters

Marshall Brian, Robert Lamb and Patrick J Kiger, 'How Nuclear Power Works', *How Stuff Works*, 31 August 2018. https://science.howstuffworks.com/nuclear-power.htm.

Review of Charles Perrow, *Normal Accidents: Living with High Risk Technologies*, 1984, by Daniel E Whitney, Massachusetts Institute of Technology, 2003.

R Smallwood BSc (Hons), MIIEE, memories of Windscale. Personal communication with author, 2020.

Jim Mahaffey, *Atomic Accidents: A History of Nuclear Meltdowns and Disasters, from the Ozark Mountains to Fukushima*. Pegasus Books, 2014.

J Samuel Walker, *Three Mile Island: A Nuclear Crisis in Historical Perspective*. University of California Press, 2006.

J Greenwood, 'Contamination of the NRU Reactor in May 1958.' Atomic Energy of Canada Limited (AECL), Chalk River Project, 1959.

'Nuclear Power in Czech Republic', World Nuclear Association, March 2020. https://www.world-nuclear.org/information-library/country-profiles/countries-a-f/czech-republic.aspx.

Mycle Schneider, 'The Reality of France's Aggressive Nuclear Power Push.' Bulletin of the Atomic Scientists, 2008. https://thebulletin.org/2008/06/the-reality-of-frances-aggressive-nuclear-power-push-2/#.

'An Investigation into a Dropped Fuel Element Incident at Chapelcross Nuclear Power Station.' A report by HM Nuclear Installations Inspectorate, Health & Safety Executive, 2002.

Available at: https://web.archive.org/web/20060625004758/
http://www.hse.gov.uk/nuclear/chapelx.pdf.

Thomas Filburn and Stephen Gregory Bullard, *Three Mile Island, Chernobyl and Fukushima: Curse of the Nuclear Genie*. Springer Books, 2016.

Chapter 20. Chernobyl – The Worst So Far

Serhii Plokhy, *Chernobyl: The History of a Nuclear Catastrophe*. Basic Books, 2018. A Chernobyl survivor and award-winning historian 'mercilessly chronicles the absurdities of the Soviet system', according to the *Wall Street Journal* review, in this well-informed account of what really went on.

'RBMK Nuclear Power Plants: Generic Safety Issues', International Atomic Energy Agency, May 1996. Available at: https://inis.iaea.org/collection/NCLCollectionStore/_Public/28/047/28047194.pdf.

Svetlana Alexievich, *Voices from Chernobyl: The Oral History of a Nuclear Disaster*. Translated from Russian by Keith Gessen. Picador, 2006. Alexievich won the Nobel Prize for Literature in 2015.

Report of the United Nations Scientific Committee on the Effects of Atomic Radiation to the General Assembly. UNSCEAR, 2008.

'INSAG-7 The Chernobyl Accident: Updating of INSAG-1', Safety Report by the International Nuclear Safety Advisory Group. IAEA, 1992. Available at: https://www-pub.iaea.org/MTCD/publications/PDF/Pub913e_web.pdf.

'Chernobyl: The True Scale of the Accident', World Health Organization (WHO), September 2005. https://www.who.int/news-room/detail/05-09-2005-chernobyl-the-true-scale-of-the-accident.

'Chernobyl's Legacy: Health, Environmental and Socio-Economic Impacts' The Chernobyl Forum: 2003–2005. IAEA, April 2006. Available at: https://www.iaea.org/sites/default/files/chernobyl.pdf.

Chapter 21: Eastern Enigmas: China, India, Pakistan – And Israel

Orde F Kittrie, 'Averting Catastrophe: Why the Nuclear Nonproliferation Treaty Is Losing Its Deterrence Capacity and How to Restore It', *Michigan Journal of International Law* 28(2), 2007. Available at: https://repository.law.umich.edu/cgi/viewcontent.cgi?article=1177&context=mjil.

Mary Beth D Nikitin, 'Comprehensive Nuclear-Test-Ban Treaty: Background and Current Developments', Congressional Research Service Report, 2016. Available at: https://fas.org/sgp/crs/nuke/RL33548.pdf.

'Congressional Research Service Reports on Nuclear Weapons.' Congressional Research Service, Federation of American Scientists, 2020. https://fas.org/sgp/crs/nuke/index.html. A useful list of all nuclear topics for use by US Congress.

'Chinese Nuclear Program', International Nuclear Programs, Atomic Heritage Foundation, 2018. https://www.atomicheritage.org/history/chinese-nuclear-program.

'India's Nuclear Weapons Program.' Nuclear Weapon Archive, 2001. https://nuclearweaponarchive.org/India/index.html.

Mary Beth Nikitin and Paul K Kerr, 'Pakistan's Nuclear Weapons', Congressional Research Service Report, 2016. Available at: https://fas.org/sgp/crs/nuke/RL34248.pdf.

Warner D Farr, 'The Third Temple's Holy of Holies: Israel's Nuclear Weapons'. The Counterproliferation Paper No. 2, Future Warfare Series No. 2, United States Air Force Counterproliferation Center, 1999. Available at: https://fas.org/nuke/guide/israel/nuke/farr.htm.

Rachel Revesz, 'Colin Powell Leaked Emails – Israel has "200 Nukes all Pointed at Iran", Former US Secretary of State Says', *The Independent*, 16 September 2016. https://www.independent.co.uk/news/world/americas/colin-powell-leaked-emails-nuclear-weapons-israel-iran-obama-deal-a7311626.html.

Paul K Kerr, 'Iran's Nuclear Program: Tehran's Compliance with International Obligations', Congressional Research Service Report, updated April 2020. Available at: https://fas.org/sgp/crs/nuke/R40094.pdf.

'Iran's Ballistic Missile and Space Launce Programs', Congressional Research Service, updated January 2020. Available at: https://fas.org/sgp/crs/nuke/IF10938.pdf.

Paul K Kerr, 'Iran's Nuclear Program: Status', Congressional Research Service Report, updated December 2019. Available at: https://fas.org/sgp/crs/nuke/RL34544.pdf.

Mary Beth D Nikitin, 'North Korea's Nuclear and Ballistic Missile Programs', Congressional Research Service, updated June 2019. Available at: https://fas.org/sgp/crs/nuke/IF10472.pdf.

Chapter 22. Japan in the Dock

'Amending Japan's Pacifist Constitution', Backgrounder, Institute for Security & Development Policy, 2018. https://isdp.eu/publication/amending-japans-pacifist-constitution/. Background brief, updated May 2020.

Michael Reilly, 'Insight: Where Not to Build Nuclear Power Stations', *New Scientist*, 25 July 2007. https://www.newscientist.com/article/mg19526144-500-insight-where-not-to-build-nuclear-power-stations/?ignored=irrelevant.

Katsuhiko Ishibashi, 'Why Worry? Japan's Nuclear Plants at Grave Risk from Quake Damage', *International Herald Tribune* and *The Asia-Pacific Journal: Japan Focus* 5(8), 2007. Available at: https://apjjf.org/-Ishibashi-Katsuhiko/2495/article.html.

'Nuclear Power in Japan', World Nuclear Association, 2020. https://www.world-nuclear.org/information-library/country-profiles/countries-g-n/japan-nuclear-power.aspx.

Kurt M Campbell, Robert J Einhorn and Mitchell B Reiss, *The Nuclear Tipping Point: Why States Reconsider Their Nuclear*

Choices. Brookings Institution Press, 2004.

Jonathon Schell, *The Seventh Decade: The New Shape of Nuclear Danger*. Metropolitan Books, 2007.

Robert Windrem, 'Japan Has Nuclear "Bomb in the Basement", and China Isn't Happy', NBC News, 11 March 2014. https://www.nbcnews.com/storyline/fukushima-anniversary/japan-has-nuclear-bomb-basement-china-isn-t-happy-n48976.

Stephanie Condon, 'Donald Trump: Japan, South Korea Might Need Nuclear Weapons', CBS News, 29 March 2016. https://www.cbsnews.com/news/donald-trump-japan-south-korea-might-need-nuclear-weapons/.

Chapter 23. Japan's Nuclear Nightmare: Fukushima

David Lochbaum, Edwin Lyman, Susan Q Stranahan, and the Union of Concerned Scientists, *Fukushima: The Story of a Nuclear Disaster*. The New Press, 2014.

Gina L Barnes, 'Origins of the Japanese Islands: The New "Big Picture"', *Japan Review* 15, 2003.

Tatsuo Usami, 'Study of Historical Earthquakes in Japan', *Bulletin of the Earthquake Research Institute* 54(3), 1979. Available at: https://repository.dl.itc.u-tokyo.ac.jp/?action=pages_view_main&active_action=repository_view_main_item_detail&item_id=33094&item_no=1&page_id=28&block_id=31.

Yoichi Funabashi and Kay Kitazawa, 'Fukushima in Review: A Complex Disaster, a Disastrous Response', *Bulletin of the Atomic Scientists* 68(2), 2012. Available at: https://journals.sagepub.com/doi/full/10.1177/0096340212440359.

The Official Report of the Fukushima Nuclear Accident Independent Investigation Commission', The National Diet of Japan, 2012.

Katsuhiko Ishibashi, 'Genpatsu-Shinsai: Catastrophic Multiple Disaster of Earthquake and Quake-induced Nuclear Accident Anticipated in the Japanese Islands'. Presented at the 23rd General Assembly of International Union of Geodesy and

Geophysics, Sapporo, Japan, 2003. Available at: https://historical.seismology.jp/ishibashi/opinion/0307IUGG_Genpatsu_Abstract.pdf.

Chapter 24. False Alarms

Max Fisher, 'Hawaii False Alarm Hints at Thin Line Between Mishap and Nuclear War', *New York Times*, 14 January 2018. https://www.nytimes.com/2018/01/14/world/asia/hawaii-false-alarm-north-korea-nuclear.html.

Kristyn Karl and Ashley Lytle, 'This is Not a Drill: Lessons from the False Hawaiian Missile Alert', *Bulletin of the Atomic Scientists*, 10 January 2019. https://thebulletin.org/2019/01/this-is-not-a-drill-lessons-from-the-false-hawaiian-missile-alert/.

Alan F Philips, '20 Mishaps That Might Have Started Accidental Nuclear War', Nuclear Files, a project of the Nuclear Age Peace Foundation, 1998. http://nuclearfiles.org/menu/key-issues/nuclear-weapons/issues/accidents/20-mishaps-maybe-caused-nuclear-war.htm.

'NORAD Agreement', North American Aerospace Defense Command, 2013. https://www.norad.mil/Newsroom/Fact-Sheets/Article-View/Article/578772/norad-agreement/. For the complete agreement see: https://2009-2017.state.gov/documents/organization/69727.pdf.

Matt Stevens and Christopher Mele, 'Causes of False Missile Alerts: The Sun, the Moon and a 46-Cent Chip', *New York Times*, 13 January 2018. https://www.nytimes.com/2018/01/13/us/false-alarm-missile-alerts.html.

'January 25, 1995 – The Norwegian Rocket Incident', History Office, United States European Command (EUCOM), 23 January 2012. https://web.archive.org/web/20160105033448/http://www.eucom.mil/media-library/article/23042/this-week-in-eucom-history-january-23-29-1995.

Sewell Chan, 'Stanislav Petrov, Soviet Officer Who Helped Avert Nuclear War, Is Dead at 77', *New York Times*, 18 September 2017.

https://www.nytimes.com/2017/09/18/world/europe/stanislav-petrov-nuclear-war-dead.html.

Chapter 25. The Sum of All Fears: Terrorists with a Nuke

M Maerli, 'The Threat of Nuclear Terrorism', International Atomic Energy Agency, 1989. Available at: https://inis.iaea.org/collection/NCLCollectionStore/_Public/32/057/32057757.pdf?r=1&r=1.

'Nuclear Terrorism: Working to Keep Nuclear Weapons Out of the Hands of Terrorists', The Nuclear Terrorism Initiative. https://www.nti.org/about/nuclear-terrorism/.

'Nuclear Power Plants: Vulnerability to Terrorist Attack.' Globalsecurity.org

Graham Allison, *Nuclear Terrorism: The Risks and Consequences of the Ultimate Disaster.* Constable, 2006.

'Terroranschlag auf Atomkraftwerk Biblis würde Berlin bedrohen', *Der Spiegel*, November 2007. https://www.spiegel.de/wissenschaft/natur/neue-studie-terroranschlag-auf-atomkraftwerk-biblis-wuerde-berlin-bedrohen-a-519668.html.

'Biblis nicht gegen Flugzeugabsturz geschützt.' *Der Spiegel*, January 2007. https://www.spiegel.de/politik/deutschland/reaktorsicherheit-biblis-nicht-gegen-flugzeugabsturz-geschuetzt-a-459184.html. *Der Spiegel* carries numerous other articles on nuclear terrorism.

Paul Reynolds, 'On the Trail of the Black Market Bombs', BBC News, 12 February 2004. http://news.bbc.co.uk/1/hi/world/americas/3481499.stm.

Alex Rodriguez, 'Cables Reveal Doubts About Pakistani Nuclear Security', *Los Angeles Times,* 29 November 2010. https://www.latimes.com/archives/la-xpm-2010-nov-29-la-fg-wikileaks-pakistan-20101130-story.html.

David Seed, *US Narratives of Nuclear Terrorism Since 9/11: Worst-Case Scenarios.* Palgrave Macmillan, 2019.

Chapter 26. Future Uncertain? Summary and Conclusions

Eric Schlosser, *Command and Control: Nuclear Weapons, the Damascus Accident, and the Illusion of Safety*. Penguin Books, 2013.

Kevin G Briggs, *Nuclear Risks and Preparedness*. CreateSpace Independent Publishing, 2014.

Michael Lewis, *The Fifth Risk: Undoing Democracy*. W. W. Norton and Company, 2019.

Toby Ord, *The Precipice: Existential Risk and the Future of Humanity*. Hachette Books, March 2020.

John Mueller, 'Nuclear Weapons Don't Matter, But Nuclear Hysteria Does', *Foreign Affairs*, November/December 2018. https://www.foreignaffairs.com/articles/2018-10-15/nuclear-weapons-dont-matter.

Jeanne Meserve, 'Mouse Click Could Plunge City into Darkness, Experts Say', CNN, 27 September 2007. https://edition.cnn.com/2007/US/09/27/power.at.risk/index.html.

Anna Saito Carson, 'Future of Warfare', Center for a New American Security Washington, DC, 2020.

INDEX

Abelson, Philip 37
Able Archer (1983) x, 223–34
A-bombs, *see* atomic bombs
Adelstein, Jake 298–300
Afghanistan 327–8, 333
al-Qa'ida xv, 328, 336–8
Albright, David 337
Aldermaston 56, 171, 188, 191, 337
alpha particles 4, 22–3
americium 18
Andropov, Yuri 224, 226, 232, 320
anti-aircraft missiles 145, 204, 216
anti-nuclear sentiment 56, 73–7, 290
anti-Semitism 5, 33
appeasement 7
Argayash incident (2017) 178
Argentina 330, 349
Arkhipov, Vasili x, 216, 220
Armenia 349
Arms Limitation Talks 229
Arsenal 31 (Russia) 54
'Asian flu' incident (1957) 56
Asos belt-stud incident (2013) 161–2
atmospheric tests 59–70, 71–7, 82–4

Atomic Age, birth of 25
Atomic Archive 288
atomic bomb disease 19
Atomic Energy Authority (AEA) 121, 169, 200
Atomic Energy of Canada Limited (AECL) 154–7
Atucha I incident (1973) 330
Australia 63, 74, 76, 296, 333
Austria 261
autism 182
Autumn Forge tests 223–4
Azores incident (1968) 92–3

Babenko, Valentin 106–7
Baikonur Cosmodrome 140, 142, 144
Ban Ki-Moon 323
Baneberry Test (1970) 81
Bangladesh 274, 349
barium 151
Becquerel, Henri 3, 19, 21
Beirut 229
Belarus 261, 262, 265, 349
Belgium 6, 196, 350

Belloc, Hilaire 43
Ben-Gurion, David 277–9
Benkö, Marietta 125
Bent Spear events 150
Beria, Lavrentiy 175
Berlin Crisis (1961) 201
Bermuda 104
beta particles 4, 23–4
Bhagavad Gita xii, 15
Bikini Atoll 60, 62–6, 67, 73, 77
bin Laden, Osama 328, 338
biological warfare xi, 53, 345–6
Bockscar 16
Boeing Corporation 144–5
Boer War 4
Bohr, Niels 4, 5, 13
Boko Haram 334
Bomarc missile incident (1960) 144–8
Boxing Day tsunami (2004) 180–1
Bradwell incident (1966) 330
Bragg, Peter N, Jr 37–8
Brazil 163–4, 166–7
Brezhnev, Leonid 143–4, 226
brinkmanship 201–2
Britain, *see* United Kingdom
Britanov, Igor 105–6
British Columbia 45, 46
broken arrow, defined 44
B-29 incident (1957) 48
B-36 incident (1957) 46–8
B-47 incident (1956) 48
B-52 incidents (1961, 1964, 1966) 127–36, 150
Bulgaria 261
Burma 12
Burtsev, Vice Adm. Oleg 120
Byelorussia 261, 262, 265, 349; *see also* Belarus

Cabon, Christine 76
cadmium 10
caesium 18, 27, 158–9, 163–5, 171, 178, 193, 197, 243, 261, 307, 338
Calder Hall 192, 193, 200, 250
CND 56
Canada 3, 44, 45, 123, 125, 154, 240, 272, 334
cancer xiii, 22, 24–6, 137, 153, 154–7, 182, 262
carbon-monoxide poisoning 54
Carnegie Endowment 270
Carp, Lt Bob 46–7, 49
carrier-killer missiles 346
Carter, James 40
Carter, Jimmy 222, 241, 245
Castle Bravo, *see* Bikini Atoll
Castro, Fidel 202, 203, 206, 210
Chadwick, James 4
Chagan Test (1965) 82–4
chain reaction 10, 13, 15, 35, 60–1, 92, 244, 258–60, 293, 312
Chalk River incidents (1952, 1958, 2008/9) 240–2
Chapelcross incidents (1969, 2001) 249–50
Charlie Hebdo 334
Chechnya 53, 338
Cheka 324
chemical warfare xi, 181, 184, 343
chemistry, study of 2–3
Chernenko, Konstantin 224, 226, 232, 233
Chernobyl ix, xv, 17, 18, 22, 25–6, 27, 84, 153, 154, 162–3, 173, 174, 178, 236, 239, 251, 253–65, 304, 305, 307, 344
Chiang Kai Shek 267
Chile 301
China 43, 76, 218, 267–71, 272, 274–5, 277, 283, 288, 316–17, 333, 346, 349
Chirac, Jacques 76
Chrome Dome incident (1968) 132
Churchill, Winston 11, 187–8
CIA 204, 225, 263, 268, 277, 316, 327, 337, 351

civil aviation disasters 226–32, 238
civil defence 228, 231, 264
civil engineering 79–84, 343
Clausewitz, Carl von 323, 345
CND 56
cobalt 27, 161, 162, 163
Cockcroft, John 187–8, 190, 193
'cold cat' incident (1965) 132–3
Cold War 70, 82, 101, 114, 125, 131, 146–7, 149, 170, 201, 212, 223–34, 242–3, 281, 313, 316, 321, 322, 329
ComPac 42
Comprehensive Nuclear Test Ban Treaty (1996) 82
Congo 3, 6
Conrad, Joseph 95
conspiracy theories 118–20
contamination 18, 48, 56, 65–6, 81–3, 131–6, 148, 164–5, 171–5, 185, 241, 243, 247, 259, 261–2
credible minimum deterrence 272
critical mass 13–14
cruise missiles 106, 121, 151, 226, 230–2, 281
C3 (Command, Control and Communications) 230
Cuban Missile Crisis (1962) x, 201–16, 313, 314, 330
Curie, Irène 24
Curie, Marie 4, 17, 20–2, 24, 29–31, 30, 154
Curie, Pierre 4, 20–1, 29–31, 32
cyber attacks 273–4, 283–4, 345
Czechoslovakia 6, 242, 261

Daghlian, Harry 33–4, 36
Daigo Fukuryū Maru, see Fifth Lucky Dragon
Das Unterseeboot (German Imperial Navy) 88
de Gaulle, Gen. Charles 71–3
De Geer, Lars-Erik 260

death ash 65
Debierne, André-Louis 21
decontamination 47–8, 131–3, 136–7, 165–6, 241, 246–8, 264–5, 282, 344
DEFCON 209, 230, 314, 318
deformation professionelle 224
Demon Core 33–5
Denmark 13, 95, 132–3, 177, 196
deuterium 61
diffusion 10–11, 37
Dimona nuclear facility 277–82
dioxin 182
dirty bombs 329, 332, 335, 337–8
'Divider' test (1992) 81–2
Dobrynin, Anatoly 208, 210–12, 224, 231, 232
Dr Strangelove 130, 218
Dōnen 292
Drebbel, Cornelis 85–6
Duluth incident (1962) 314
Dungeness 171
DuPont incidents 178–9
Dyatlov, Anatoly 257
Dzerzhinsky, Felix 324

E I du Pont de Nemours & Company 179
earthquakes 235, 276, 288, 290, 296–302, 306, 345
East Germany 228
EBR-I incident (1955) 239–40
EDF incidents (2004, 2009, 2011) 248–9
Eduarda, Maria 166–7
Egypt 311, 317–18, 326
Einstein, Albert 4–9, 29, 50–1
Eisenhower, Dwight D ('Ike') 72
Eisenhower, Mamie 90
electromagnetics 15, 24, 70
electrons, discovery of 4
engineering paper (1946) 79–80
Enola Gay 16
Enrico Fermi Unit incident (1966) 242

Enterprise, USS ('Big E') 90
ExComm 205–6, 209, 210, 212

Fabius, Laurent 75–6
Fairley, Peter 271
fallout, first coined 65
fast breeder reactors 242, 272
Fat Girl 115
Fat Man 16, 39
fault lines 276, 296, 326
Fermi, Enrico 4–5, 10, 15, 60, 242
Feynman, Richard P xvii
Fifth Lucky Dragon 64–6
Finland 51, 350
Flight 007 incident 226–7, 229, 230, 232, 320
Flyorov, Georgy 50–1
Formosa, *see* Taiwan
Four-Pillars Nuclear Policy 289–90
France 71–7, 92, 151, 247, 261, 284, 288, 334, 339, 349, 350
Free French 72
Frisch, Otto 5–6, 33
Fukushima ix, 22, 154, 170, 178, 235–6, 269–71, 275, 276, 281, 288, 290, 291, 293, 295–308, 344–5
Fulbright, J William 276–7
fundamentalism xv, 325–9, 333–8
fusion bomb 60

gamma particles 3, 24
Gauvenet, Andrá 247–8
GCHQ 231
gene damage 25
Germany 3, 196, 218, 261, 268, 284, 339, 350
anti-Semitism in 5, 33
Gilpatric Report 280
Glasnost 109, 120
Global Initiative to Combat Nuclear Terrorism 331–2
Global Research 302, 307–8

Goiânia incident (1985) 163–5
Goldsboro incident (1951) 127–31
Gorbachev, Mikhail 233–4, 253
Gorbachev, Raisa 233
Gordievsky, Col Oleg 224, 230, 231
graphite 10, 193, 199, 247, 254, 258, 258–9, 259
Great Kanto earthquake (1923) 296
'Great Patriotic War' (WWII) 50
Greece 261
Greenham Common 56, 226
Greenland 99, 132–3, 216, 316
Greenpeace 74–6, 76, 180
Gromyko, Andrei 207, 228, 233
groundwater contamination 148, 165, 170, 171, 262
Groves, Lt-Gen. Leslie R 9, 11, 38
GRU 23, 202
Guam 39–40, 48
Guantánamo Bay 207
Gunpowder Plot (1605) 325

H-bombs, *see* hydrogen bombs
half-life, first coined 4
Hamaoka incident (1981) 292
Hanford Site 11, 25–6, 169–70
Hao Atoll 76
Haq International 161
Harwell 187–8, 190
Hawaii 70, 301, 309–11
health problems 19, 25–7, 30–2, 136–7, 153–67, 345
heavy water 12
hibakusha survivors 65
Hicks, Bob 148–50
high-altitude tests 70
Himalayan War (1962) 272
Himmler, Heinrich 33
Hinton, Christopher 188, 190–1
Hiroshima xi, xvi, 16, 18, 22, 25, 26, 33, 39, 51, 52, 54, 63–5, 68–9, 129, 131, 135, 153, 175, 219, 260, 287

Hitler, Adolf xiv, 5, 6, 11–12, 343
Hitzfelder, David 90–1
human error 32, 48–9, 118, 169, 199, 243, 254, 257, 291, 310–11, 315, 322
hybrid warfare 284, 345
Hyde Park Agreement (1944) 11
hydrogen bomb 60, 191, 202
hydrogen peroxide (HP) 114
hypergolic fuel 139

ICBMs (intercontinental ballistic missiles) 140, 141, 148, 149, 183, 205, 209, 215–16, 221, 269, 281, 313, 314–15, 319–20
IEDs (improvised explosive devices) xii
igloo (storage bunker) 48, 50
India 63, 151, 161, 267, 271–4, 280, 325, 349, 350
Indian Strategic Nuclear Command (SNC) 272
Indianapolis incident (1945) 38–42
Indo-China 268
induced radioactivity 5, 24–5, 163
infertility 182
Institute of Radiation Medicine and Ecology (IRME) 173–5
International Atomic Energy Agency (IAEA) 121, 159–60, 163, 236, 251, 276, 281
International Commission for Radiation Protection (ICRP) 26, 174
International Nuclear and Radiological Event Scale (INES) 163, 171, 172, 178, 223–34, 235–6, 243, 247, 251
iodine 27, 193, 197, 199, 272
Iran 282, 283–4, 287, 326, 327, 350
Iraq 283, 329, 333, 338
ISIS/Islamic State xv, 284, 329, 333, 334, 337–8
isotope separation 9–10
Israel 76, 92, 177, 267, 276–85, 287, 318, 349
Italy 5, 116, 179–82, 182, 213, 261

Japan 15, 124, 269–71, 341, 349, 350
A-bombs dropped, *see* Hiroshima; Nagasaki
Fukushima, *see main entry*
safety agencies 291, 292, 304, 305, 306
'safety' record 287–93
Jaslovské Bohunice incident (1977) 242–3
'Jericho' series 281
jihadi groups, *see* fundamentalism
Johnson, Lyndon B ('LBJ') 280, 288
Jones, Parker F 129–30

Kalpakkam incident (2003) 272–3
Kanupp incident (2011) 275
Karaul incident (1953) 174
Kazakhstan 51, 82, 140, 173–5, 349
Kennedy, John F ('JFK') ix, x, 201–15 *passim*, 276–9
Kennedy, Robert F ('RFK') 206, 210, 211
Kenya 165–6
KGB 13, 119, 209, 210, 224–31, 321
Khrushchev, Nikita 67, 101, 143, 202–15 *passim*
Kim Jong-un 285, 309
Kinchole incident (1973) 318
Kissinger, Henry 316, 318, 346
Klebanov, Ilya 118, 119
K-19 incidents 54–5, 95, 96, 98–100, 216
Kohl, Helmut 230
Korchilov, Lt Boris 99–100
Korea 12, 226–7, 229, 230, 232, 268, 271, 284, 309, 320, 350
Korean War 268
Kosmos 954 123–6
Kramatorsk incident (1980s) 158–9
Kramish, Arnold 37–8
Kudankulam incident (2019) 273–4
Kursk 53, 113–22
Kyshtym disaster (1957) 176–8

La-Z-Boy recliner incident (1998) 162
lanthanum xx, 151
Latvia 261
Lavoisier, Antoine 1–2
Lebanon 282
Leonov, Capt. Pavel 101, 102
leukaemia xiii, 158, 159, 166
leukopenia 137
Lima-02 incident (1964) 148–50
Limited Nuclear Test Ban Treaty (1963) 70, 83
liquefied natural gas (LNG) 288
lithium 60, 61, 63, 191
Lithuania 261
Little Boy 16, 18–19, 38–9, 68
Litvinenko, Alexander 23
Livingston, David 139
L-IV incident (1942) 32–3
'lone wolf' terrorists 332, 338–9
'loose nukes' 338
Los Alamos 11, 13, 15, 33, 177, 188 (*see also* Manhattan Project)
loss-of-coolant accidents 103
Lucens incident (1969) 246–7
Lyudmila incident (1998) 178

McCormick, Maj. Thomas W 131–2
Macmillan, Harold 199, 200
McNamara, Robert 206, 217, 221
McNeill, David 298–300
McVay, Capt. Charles, III 38–40
magnesium 126, 191, 247
Magnox 249, 250–1, 289
Manchester bombing (2017) 338
Manchuria 13, 268
Mancuso, Prof. T F 25–6
Manhattan Project xiv, 8–11, 25–6, 33–9, 51, 59, 60, 153
 first nuclear explosion 14–16
Mao Zedong 267–8, 316–17
March 1940 memorandum (Peierls, Frisch) 5–6

Mariana Islands 14, 38–9
Marshall Islands 60, 62–3, 66, 73
maskirovka operation (1962) 216
Mattocks, Lt Adam 127, 128
MAUD 5
Mayak facility 175–9
Mazurenko, Vyacheslav 101–3
Medvedev, Grigori 255–6
Medvedev, Zhores 177, 178
Meigs, Douglas P 37–8
Mengele, Dr Josef 59
Messinger, Maj. Larry 133–4
Metropolitan Edison 244–5
Mihama incident (2004) 293
milk sampling 197–8, 199
Ministry of Defence (MOD) (UK) 55, 56–7
Minuteman incidents (1983, 2014) x, 150, 319–21
MI6, *see* SIS
MIT studies 238–9
molecular damage 25
Moon landings 81, 183
Moorestown incident (1962) 314–15
Mororua Atoll 73, 74, 75
Moscow Metro bombings (2010) 332
Mowatt-Larssen, Rolf 336–7, 339
MRBMs (medium-range ballistic missiles) 204
Mrima Hill 165–6
Multan meeting (1972) 274
Muslim Brotherhood 326
Mussolini, Benito 33

Nagasaki xi, 16, 22, 25, 26, 39, 51, 65, 69, 129, 131, 153, 175, 260, 287
Nakajimab, Stephanie 299
Narora incident (1993) 272
NASA 142, 144
national security, catch-all cloak of 57
NATO 73, 106, 109, 115, 116, 118, 133, 151, 202, 211–12, 219, 223–32, 342

natural uranium, *see* U-238 *under* uranium
Nautilus, USS 89–90, 98
Navy Yard incident (1944) 37–8
'Ndrangheta 179–82
Nedelin Incident (1960) 140–4
Nehru, Jawaharlal 272
NEPTUN 10P 157–8
neptunium 18
nerve agents 343
Netanyahu, Benjamin 281
Netherlands 85, 121, 261, 350
neutron bombardment 4, 5, 10
Nevada Test Site 80, 81
New Zealand 74–6, 76, 77
Newfoundland 46
Nigeria 333, 334
NIMBY-ism 270–1
9/11 terrorist attacks on the USA xv, 325, 328
nitric acid 3
Nixon, Richard 318
NKVD 13, 175
no-first-use policy 268, 269, 272
Nobel Prizes 5, 30, 33
Non-Proliferation Treaty (NPT) 267, 280, 280–1, 281, 282, 289, 290
NORAD 124, 125, 313, 314–15, 318
NORAD incident (1979) 318–19
North Korea 284, 309, 349
North Pole 90, 124, 225, 313
Northern Lights incident (1962) 313–14
Norway 12, 106–9, 118, 196, 301, 313
Novaya Zemlya 68
NSA 231
nuclear accidents, seven levels of (INES scale) 236–7
Nuclear Consulting Group 172
nuclear fallout, *see* fallout
nuclear first strike 203, 210, 212, 228, 229, 231–2
nuclear fission 6, 9, 12–13, 27, 50

Nuclear Free Zones 75
Nuclear Information Service (NIS) 55
Nuclear Materials and Equipment Corporation (NUMEC) 278
nuclear meltdowns 99, 216, 239–40, 337
nuclear proliferation xv, 71, 76
nuclear radiation, *see under* radiation/radioactivity
Nuclear Regulatory Commission (NRC) 246
Nuclear Test Ban Treaties 70, 72, 76
nuclear waste, *see* waste disposal
Nuffield Institute 25
Nyonoksa incident (2019) 151

Oak Ridge incident (1972) 330
Obama, Barack 246
Office for Nuclear Regulation (ONR) 170–3
Ohi incident (2011) 305
Oopa, Pouvanaa a 73–4
Operation Crossroads 62
Operation Dropshot 43
Operation Kama 216
Operation Moist Mop 136
Operation Olympic xiv
Opération Satanique 75
Oppenheimer, J Robert xiv, 11, 15, 60, 329
Orange Herald test 191
organised crime 179–82, 204–5
Organization for Economic Co-operation and Development/Nuclear Energy Agency (OECD/NEA) 236
'orphaned' radioactive sources 161–2

pacifism 7, 9, 289
Pakistan 151, 267, 274–6, 280, 333–8, 349, 350
Palestine 277, 324, 326
Panama 86

PAVE PAWS 318–19
'peaceful' nuclear explosions 80–1
Pearl Harbor 9, 39, 205
Peierls, Rudolf 5–6
Penney, Prof. Sir William 56, 188, 198–200, 351
Percy, Charles 318
Pereira, Fernando 75, 77
periodic table 3
Permissive Action Links (PAL) 221
Perrow, Charles 237–8
Pershing missiles 229, 230, 231
Petrov, Lt-Col Stanislav x, 320
PH241 Nuclear Studies programme 292–3
Philippines 39
pitchblende (uraninite) 3, 6, 20–1
'Playing With Fire' (NIS) 55
plutonium 18, 25, 34–6, 44–5, 169, 178, 190, 191, 193, 250, 272, 283
discovery of 10
injections 59
Poland 8, 261
Polaris 205
polonium 20, 23
discovery of 29
Polynesia 73–7
Popov, Adm. 116, 118
potassium 159, 165
Potsdam Conference (1945) 51
practice depth charges (PDCs) 217, 219
Preminin, Sergei 105
Price, Terence 190
Prieur, Dominique 76
Project 641s (Soviet Russian submarines) 218
Project Plowshare 80–1
propaganda 67, 146, 177, 213, 226
PSYOPS 224–5, 231
Pushkin, Alexander 101
Putin, Vladimir 117, 119, 121

Quebec 46
Quebec Conference/Agreement (1943) 11
radiation/radioactivity 17–27
Curies study 29–30
described xiii–xiv
early USSR research 50
measurements for 19–20
radiation sickness, *see main entry*
radioactive contamination 18
study of 3–4
radiation sickness 19, 21–7, 30–2, 136–7, 153–67, 173
radiation therapy 153
radioactive contamination, *see* contamination
radioactive fallout, *see* fallout
radioactive waste, *see* waste disposal
radiotherapy 154–7, 161–7
radium 3, 32
discovery of 21–2, 29
Rainbow Warrior 75–7
Rarotonga, Treaty of 75
RBMK reactor incident (1986), *see* Chernobyl
reactor scram 92
Reagan, Ronald 202, 224, 226, 228, 230–3, 321
Reagan's 'bombing USSR in five minutes' incident (1984) 321–2
Reforger 83 223
religious prophecies xii, xiv, 15
Rickover, Adm. Hyman 85, 88–90, 91
Ring of Fire (Pacific) 296
Rising Moon false alarm (1960) 313
Romania 261, 350
Rongelap Island 66
Röntgen, Wilhelm 3, 19–20
Roosevelt, Franklin D ('FDR') 6–9, 11, 50–1
Royal Navy (RN) 86, 87, 114, 118
Royal New Zealand Navy (RNZN) 74

R-16 incident 140–4
RTG incident (2001) 159–61
Rudd, Kevin 267
Runermark, Claes-Göran 261
Rupp, Rainer 228–30
Russian Federation 117, 119, 121, 284, 349 ; see also USSR
Rutherford, Ernest 4

Saddam Hussein 283
Saint-Cyr Military Academy 72
Saint-Laurent incidents (1969, 1980) 247–8
Saladin (Salah ad-Din) 326
Samson Option 285
Samut Prakan incident (2000) 162
San Andreas Fault 296
San Cristóbal 216
Santa Susana Field Laboratory 184–5
Satō, Eisaku 288, 289
satellites 70, 118, 123–6, 160, 226, 242, 262–3, 269, 280, 313–16, 319–21
Savannah River spillages 178–9
Savitsky, Capt. Valentin 218, 219–21
Schreiber, Raemer 35–6
scrap radioactive metal, recycling of 161–3
Seascale, see Windscale
Sedan Crater 80–1
seismology 80
self-defence policy 288, 289
self-denying ordinance 343
Sellafield 171, 200
Semipalatinsk Test Site 51, 82, 173–5
Shahzad, Faisal 333
SHAPE x, 223
Shapiro, Zalman 278
Siberia 54, 210, 316–17
sicarii 324
Sicily 325
SIS (MI6) 228
SISMI 180

Skaar, Victor 136–7
Skylark, USS 91–2
Slaughter, Ensign Gary 220–1
SLBMs (submarine-launched ballistic missiles) 269
Slotin, Louis 34–6
Smiling Buddha test (1974) 274
Socatri incident (2008) 248
Sodium Reactor Experiment (SRE) 183–5
Somalia 180–2
Sorensen, Theodore 210, 211
South Africa 280, 331, 349, 350
South Pacific Nuclear Free Zone 75
Sovacool, Benjamin 172–3
Soviet Union, see USSR
Space Law in the United Nations (Benkö) 125
Space Liability Convention 125
space probes 160
Spain 134–7, 339, 350
'special relationship', (US-UK) first coined 11
Speedlight 69
Speer, Albert 11
Spitzbergen incident 322
spy planes 204, 210, 216, 313
SRE incident (1959) 183–5
Stalin, Joseph xiv, 50, 51, 175
Starfish Prime 70
Strategic Arms Reduction Treaty (1991) 150
strontium 18, 27, 151, 160, 178
'stud 16' 227
Stuxnet 283
Submarine Surfacing and Identification Procedures 217
Submarine Thermal Reactor (STR) 89
submarines 85–93, 205, 216–19
disappearances 92–3
first-ever 85–7
first nuclear-powered 89–90

'Holland' boats 86
K-class 53–5, 92, 95–112
Oscar class 113–22
peroxide powered 114
submerged propulsion 37
Sudan 4
Suez Crisis (1956) 311, 312
Sugaoka, Kei 306–7
'sun flare' incident (1967) 315–16
Suvorov, Nikolay 109–11, 110
SVR 23
Sweden 260, 261, 350
Sweeney, Charles W 16
Switzerland 76, 180, 246, 261, 350
Syria 312, 318, 329, 334, 338
Szilárd, Leó 5, 6

Tahiti 74
Taiwan 181, 267, 269, 350
Taliban 333, 336
Tarapur incident (1989) 272
Technetium-99 170
Teller, Edward 5, 60, 62
TEPCO 297, 301–3, 306–7
Tepojaco incident (2013) 163
terrorism xv, 75, 323–39, 347
Tesla, Nikola 20
Thailand 162
Thant, U 212–13
Thatcher, Margaret 230, 232, 233
Therac-25 154–7
thermal diffusion 37
Thirteen Days 210, 211
Thor missile launch (1962) 70
thorium 165, 166
Thornburgh, Dick 245
Three Mile Island ix, 239, 244–6, 344
Three Non-Nuclear Principles 289
Thresher, USS 91–2
Thugs of India 325
thyroid 27, 175
Tibbets, Col Paul W 14, 16

Time Magazine 81, 90, 227
The Times 171, 172, 177, 321
Tinian Island 14, 38–9
Titan-II incident (1980) 139–40
TNT xii, xv, 6, 15, 67
Tokaimura incidents (1981, 1997, 1999) 292–3
Tomahawk missiles 226
'Topaz' (East German agent) 228–30
Totskoye incident (1954) 51–2
trafficking (of nuclear material) 332
Triad weapon mix 269, 272, 281
Trinity 15, 24, 38
tritium 60, 61, 171, 190, 191, 197, 242, 250
Truman, Harry S xiv, 51, 268
Trump, Donald 284
Tsar Bomba test 69
Tsugura incidents (1975, 1981) 292
tsunamis 180–1, 235, 290, 291, 297–302, 308, 345
Tsuzuki, Dr Matsuo 19
Tuamotu Archipelago 73
Tube Alloys 6, 9, 188
Tulloch, Maj. Walter 128–9
Tumerman, Lev 177
tungsten carbide 33
tungsten-carbide brick incident (1945) 33–6
Turkey 203, 211–12, 312, 350

UKAEA 198, 199
Ukraine ix, xv, 17, 18, 158, 202, 236, 253–65, 344, 349, 350
underground storage 148–9, 176, 269, 345
underground testing 80–3
United Kingdom:
accidental atomic discharges 49–50
America's 'special relationship' with 11–12
armed forces 79
cover-ups 199–200

nuclear incidents 170–3
post-war nuclear programme 188, 249–50
Public Record Office (PRO) 56
secret nuclear site 249–50
United Nations (UN) 76, 180, 212, 255, 264, 274, 283, 329
UN Development Programme (UNDP) 180
UN International Convention for the Suppression of Acts of Nuclear Terrorism 329
UN Security Council 274, 283
United States:
armed forces, *see* by name *under* US
A-bombs dropped by, *see under* Japan
civil-engineering programme 80–1, 90
Mancuso-Stewart study 25–6
'missile gap' 205
nuclear stockpile 43
post-war nuclear programme 62–6, 201–14
UK's 'special relationship' with 11–12
wartime nuclear programmes, *see* Manhattan Project; US Navy
'unknown' nuclear crisis (1969) 316–17
uranium 12, 20, 33, 37, 51, 165–6, 170, 171, 190, 247, 284, 329, 331, 349
depleted 48, 250
discovery of 3
U-235 6, 9, 10, 11, 12, 13–14, 16, 27, 56, 89, 125, 293
U-238 9–10, 12, 46, 48
'yellowcake' powder 344
Zalman Shapiro 'loses' 278
uranium hexafluoride 37
uranyl nitrate 293
Urazoko Bay incident (1981) 292
US Air Force (USAF) 44–5, 126–9, 139–40, 203, 206
accidental atomic discharges 44–9
incidents (1950s) 44–9

safety record 48, 127–36, 144–50
Strategic Air Command (SAC) 43, 44, 127, 132–3, 139, 209, 221, 268
US Army 15, 34
US Army Air Forces (USAAF) 12, 14
US Army Corps of Engineers 9
US Atomic Energy Commission (AEC) 59
US Code of Military Justice 43
US Department of Defense 46
US Department of Energy 179, 184, 332
US Department of Homeland Security 162
US Department of Veteran Affairs 137
US–Iran deal 284
US Marines 202
US National Defense Research Council (NDRC) 8–9
US National Reconnaissance Office (NRO) 263
US National Security Council (NSC) 124–5, 205
US Naval Academy 88
US Naval Court of Inquiry 92, 93
US Navy 36–9, 62, 97, 136, 216, 216–19, 241, 245
atomic development programme 37–40, 205
Cuba blockade, *see* Cuban Missile Crisis
safety record 90–3, 132–3
submarine service 86–93
US Radium Corporation 22
US Special National Intelligence Estimate (SNIE) 229
US Strategic Air Command (SAC) 221, 268
USSR:
Chernobyl, *see main entry*
civil-engineering programme 82–4
collapse of 144, 178–9
naval safety record 92, 95–112, 113–22

post-war nuclear programme 67, 175, 188, 201–14, 202, 215–22
'principal adversary' 202
wartime nuclear programme 13, 50–2
see also Russian Federation
Ust-Kamenogorsk incident (1956) 173–5
Ustinov, Vladimir 120
UXBs (unexploded bombs) 123–37

Vanin, Capt. Yevgeniy 106–8
Vanunu, Mordechai 281
Vela incident (1979) 280
Vichy Regime 72
volcanic activity 296

Warsaw Pact 208, 231
waste disposal 169–85
Watergate 318
weapons-grade uranium, see U-235 under uranium
WE177 incidents (1976, 1988) 151
West Germany 202, 254
Westinghouse Electric Corporation (WEC) 89
White Sands Missile Range 15, 24
Wigner releases 191, 192
Willsher, Kim 262, 263
Windscale incident (1957) 185, 187–200, 250, 305
Woodcock, John 342
Woolsey wildfire (2018) 184–5
World Institute for Nuclear Security (WINS) 331

World Trade Center (WTC) xv, 325, 328; see also 9/11
World War I 343
Lusitania attack 87
outbreak of 22
World War II xiv, xv, xvi, 50, 72, 87, 88, 114, 118, 170, 238, 268, 288, 343
A-bombs deployed in, see under Japan
Germany defeated 14
Japan defeated 16
Nazi nuclear programme 6–8, 11–12, 32–3
outbreak of 8
'special relationship' 11, 187–8
World War III 215–22, 312

X-rays 61, 344
discovery of 3–4, 19–21

Yarborough, Katie 155–6
'yellowcake' powder 344
Yeltsin, Boris 322
Yerofeyev, Rear Adm. Oleg 109, 111
Yom Kippur War 317–18
Yomiuri Shimbun 64
Yoshida, Masao 302–3
YouTube 67
Yucca Flat Test (1970) 80, 81
Yugoslavia 261
Yuvchenko, Alexander 259–60

Zateyev, Capt. Nikolai 98–100, 100
Znamenka incident (1956) 174